The Paraset Radio

The Story of a WWII Spy-Radio and How to Build a Working Replica

By Hiroki Kato, AH6CY

Published by

Radio Society of Great Britain of 3 Abbey Court, Priory Business Park, Bedford
MK44 3WH, United Kingdom
www.rsgb.org

First Printed 2020

ISBN: 9781 9101 9395 2

Editing: Philip Lawson, G4FCL
Production: Mark Allgar, M1MPA
Typography and design: Chris Danby, G0DWV
Cover design: Kevin Williams, M6CYB

Printed in Great Britain by CPI Antony Rowe of Chippenham, Wiltshire

Any amendments or updates to this book can be found at:
www.rsgb.org/booksextra

Contents

Preface

This book describes the gripping story behind the 'Paraset' – a unique spy-radio, dropped behind enemy lines in the dark days of WWII. Providing vital communications, SOE agents operating the Paraset put themselves in danger of discovery each time they pressed their Morse key, with many paying the ultimate sacrifice. Today, as we celebrate the 75th anniversary of the end of the war, this book not only pays tribute to the Paraset and its wartime operators, but gives you practical instructions on how to build a replica, so that you can get a taste of wartime operating, but without the danger of a fatal 'knock on the door'.

It is the author's sincere hope that many will have better appreciation for the important role played by the Paraset and other spy radios during WWII as a result of reading this book. He also hopes that more people will be building their own Paraset replicas to keep the legacy of this unique radio alive.

Foreword

I came across this radio for the first time about fifteen years ago and instantly fell in love with it, because it employs a 6V6 as its output valve; the same valve I used to build my first transmitter in 1957 when I was a newly licensed 15 year-old Amateur. It was like finding a long-lost friend.

My affection for the Paraset has deepened over the years as I have learned more and more about it. I have visited many museums in Europe looking for surviving sets and have researched how it and other spy radios were deployed by the agents and local resistance fighters "in the field," that is, in Nazi-occupied Europe. I have built three working replicas and have written a few articles on the experience for Amateur radio magazines [1]. A French Amateur, Jean Claude Buffet, F6EJU, and I have been serving as a volunteer global clearinghouse for information on all surviving Parasets and replicas built by Radio Amateurs. We are hopeful that we will be unearthing more original sets, and that enthusiasm for building replicas will continue as more people find out about this radio's unique legacy.

See you on the air.

Hiroki Kato

[1] 'The Paraset: A WWII Spy Radio You Can Build', *CQ*, Feb. 2016, pp 19-25;
'Two Clandestine Radios of WWII: Replicating the Prison Camp Radio and the Paraset Spy Transceiver', *Electric Radio* Nov. 2012, pp 26-39;
'Two Clandestine Radios of WWII- Update', *Electric Radio* March 13, 2013, pp 18-26.

Introduction

THIS IS the story of a WWII spy radio, affectionately known as the 'Paraset' by those who built and used it during the war, and later by radio enthusiasts around the world, including many Radio Amateurs [1].

The Paraset is a unique radio. Not because it was state-of-the-art at the time, and it was not the only spy-radio produced during WWII, but because it was a minimalist device designed to do a specific job - a big job as we shall see - with the fewest features and components, and the minimum necessary output power. It has no meters, volume control, or VFO; it does not even have a power switch, and only produces a maximum power output of four to five watts in CW. It is also not the easiest radio to operate, but did the intended job very well indeed, having the smallest and lightest package of all the transceivers of WWII sending clandestine signals from the European continent to England, and receiving signals sent from England to the continent.

Some 800 to 1,000 units were built by hand, although the exact number has never been disclosed in the official record. They were first made in workshops at Whaddon Hall and then at Little Horwood, near Bletchley Park, about an hour's train ride from London. They were later made in higher quantities at outside manufacturing locations [2]. Sadly, most were destroyed at the end of the war and today only a few original Parasets remain [3]. At the time of writing, 19 have been identified (see Appendix 2), mostly in museums across Europe, with a few in the hands of private collectors.

However, inspired by the Paraset's history and epic stories of bravery and sacrifice of its operators, some of which are related here, many Amateurs in Europe and North America have produced replicas and have shared their experiences on the Internet and directly with the author. With the help of Jean Claude F6EJU, I have identified at least 50 working reproductions, with varying degrees of faithfulness to the original, and the number is climbing.

This book presents instructions on how to build a working replica of the Paraset. Based on fellow builders' and the author's own experiences, practical ways of acquiring and fashioning the necessary components, or fabricating parts to the original specifications using modern materials will be described. Also discussed is the joy of operating replica sets on the Amateur bands, including actual examples of Paraset-to-Paraset QSOs. In addition, a few modern "reincarnations" of the Paraset, built with 21st century electronic components, including transistors and ICs, will be mentioned. These sets preserve the original spirit of a minimalist approach to radio building as they employ a regenerative receiving circuit rather than the now common super-heterodyne or direct conversion approach, and use a fixed-frequency crystal oscillator instead of a VFO for the transmitter.

References

[1] The origin of the moniker 'Paraset' is not definitively known. Some suspect that the affectionate reference was first made among agents of BCRA (the Bureau Central de Renseignements et d'Action - the French intelligence service located in London) deployed in France. The radio's official designation is the Whaddon Mk VII. It was most often dropped by parachute into Nazi-occupied western European countries such as France, the Netherlands and Norway for use by the resistance fighters, and SOE (the Special Operations Executive) and MI6 agents, many of whom parachuted into those countries.

[2] From personal correspondence with Geoffrey Pidgeon, the youngest of the group who worked at Whaddon Hall building the Paraset. He relates: "At first at Whaddon Hall we made them in lots of 50 as we were still making the wretched MkV (another spy radio much heavier and bulkier than the Paraset). However, in the new 'Factory' at nearby Little Horwood we started in 100s at a time, but then they pushed it up to 200." Pidgeon is aged 93 at the time of writing, and is the only known living member of the team directly involved in manufacturing the Paraset. He has been a generous correspondent over the years and I finally had the privilege of meeting him in person on the Bletchley Park Veterans Reunion Day on Sept. 2, 2018. I would like to thank him for all the information and experience he has shared with me.

Pic Intro1 Geoffrey Pidgeon and Author

[3] Pidgeon was posted to Calcutta, India, toward the end of WWII. He told how he and his fellow soldiers were ordered to dispose of 12 Parasets and other radios flown back from Kunming, China, right after the war's end. Desperate to find ways to get rid of the radios they decided to sink them in the middle of Lake Dhakuria in Calcutta. He kept one of them, which he gave to Bletchley Park. The reason why the Parasets were destroyed, and so few are seen today, has been speculated on variously. Some say it was the onset of the Cold War that prompted Churchill to order their destruction as he didn't want the Soviets to get hold of any British spy radios. Another view is that Churchill wanted to keep anything related to the SOE secret, even after the war, including the radios used by its agents. Dave Gordon-Smith, G6UUD, who has conducted extensive research into WWII spy-radio history, thinks the real reason may be more mundane - reluctance by the various armed services to pay for repatriation of items that had been taken abroad, but were no longer needed.

Dedication

"One cannot but admire without reservation the secret agents who dared to tap out messages from enemy-occupied territory or, as in many occupied countries, secretly built equipment for others to use, and those who sheltered or assisted them. We remain in their debt." - Geoffrey Pidgeon, *The Secret Wireless War: The story of MI6 Communications 1939-1945,* Arundel Books, p.97, quoting Pat Hawker G3VA.

Acknowledgments

Many Radio Amateurs have contributed to the preparation of this book by sharing their experiences of building and operating Paraset replicas. My heartfelt thanks in particular to: Jerry Fuller, W6JRY; Steve McDonald, VE7SL; Dan Peterson, W7OIL; Paul Signorelli, W0RW; Mike Murphy, WU2D; Johnny Apell, SM7UCZ; Jean Claude Buffet, F6EJU; Peter Jensen, VK7AKJ; Michael Tyler, WA8YWO; and Graham Stannett, G4VUX. In addition, spy-radio expert Henk van Zahm of the Netherlands very kindly shared much information about building replicas.

Less obvious, but equally important in the making of this book, is that many Radio Amateurs and non-Radio Amateurs read all or part of the draft and gave me valuable comment. As I am not a native speaker of English, this book has benefited to an inordinate degree from stylistic and literary refinement offered by those people. I owe a debt of gratitude to: John Swartz, WA9AQN; David Varn, KM6RI; Eric Norris, WD6DBM and his wife Cindy Martin; Bob Mix, KF6ABC; Larry Pinto, N9FIA; James Sanford; and my wife Joan Stern Kato, who not only critically read the book, but also indulged me the many hours I spent in my shack and on research travels.

Last, but most certainly not least, I owe a major debt of gratitude to Phil Lawson G4FCL, my editor at the RSGB, who not only helped reshape and refine my original manuscript into a more readable form, but has also been my most ardent champion of the project.

1

Historical Background

TO FULLY APPRECIATE the unique position that the Paraset occupied in the annals of clandestine communication devices, we need first to look at the historical background in which it was developed and deployed. The Paraset was conceived and built in the dire circumstances of WWII. In particular, the story of this and other spy radios is inseparable from the story of a secret paramilitary organization urgently created by the British government to face the grave condition of the war.

1.1 Invasion of Europe

Germany invaded Poland on September 1, 1939, without a formal declaration of war. The Nazi advance into Poland was swift and massive; the ill-equipped and ill-prepared Polish army was utterly helpless against the German onslaught. Britain and France declared war against Germany two days later, but the Allied Forces' counter-offensive was ineffective. Within less than a year Germany began aggression against surrounding countries. Belgium surrendered on May 28, 1940, Norway succumbed on June 9, and France fell on June 17. The Battle of Britain began on July 10, 1940. London was under heavy Luftwaffe bombardment night after night thereafter, for fifty-seven consecutive nights. The dire circumstances of this period were vividly depicted in two recent films: *Dunkirk,* which shows how close the British and Allied forces came to near total demise in the spring of 1940 narrowly managing to evacuate from the continent; and *Darkest Hour*, which portrays the agony that Winston Churchill suffered in those tenebrous times.

1.2 Creation of the SOE

American participation in the war was still more than a year away despite Churchill's desperate pleas to US President Franklin Roosevelt. The reality was, to many Britons, and to the people of then Nazi-occupied European countries, that in the spring and summer of 1940, things began to look hopeless. Churchill had recently become Prime Minister, and in July 1940, in an effort to try and turn the situation around, he ordered the creation of a new secret paramilitary organization called the 'Special Operations Executive' or SOE. The existence of SOE was one of the most closely guarded secrets of the war, and on a par with ULTRA [1]. Indeed, most Britons did not know anything about the SOE until many decades after WWII, as most British government records relating to the SOE and its agents were classified until the 1990s, and some remain classified even now. Many important records – up to 80%, according to the Imperial War Museum - were destroyed deliberately and by fire. The SOE was called "Churchill's secret weapon" by the small number of government and military personnel who were directly involved with it, as well as "the ministry of ungentlemanly war" and "the Baker Street Irregulars" after the location of SOE headquarters in London [2]. Since many British political and military leaders had an upper-class background, they considered the SOE to be an unsavory, undesirable, and indeed an "ungentlemanly", organization, which reflects a tension between expected standards of conduct, and the imperative of war. However, Churchill was a most practical and pragmatic leader, even though he, too, was from the upper class of British society.

According to its founding charter, SOE was tasked "to co-ordinate all action, by way of subversion and sabotage, against the enemy overseas", and hence its agents worked in close collaboration with local resistance groups behind enemy lines in the Nazi-occupied countries [3]. They were tasked with arranging such things as the drop-off of weapons, currencies, fake IDs, ration cards and the like, as well as establishing and maintaining escape routes for downed Allied pilots and important anti-Nazi political figures. When SOE was created, the areas "behind enemy lines" were all on the European continent, but its planners foresaw possible action and subversion in the event of a German occupation of Britain itself. Later, SOE also became engaged in North Africa and in Japanese-occupied areas of Asia.

The exact number of people who worked for SOE has never been revealed. M.R.D. Foot, its official historian, wrote that in 1944, at the height of operations, nearly 10,000 men and about 3,200 women had been recruited. Five thousand of these agents were either in the field, or ready to deploy to the field for work behind enemy lines. The rest were engaged in planning, intelligence, supply, research, transport and radio operations at home [4]. Some worked in four receiving stations in Britain, listening for incoming signals from agents in the field, and some as cypher clerks, encoding outgoing messages to those agents and decoding incoming messages from them. These operations required sufficient staff to be able to operate around the clock (Pic 1.1).

Pic 1.1 Home Station Receiving Room STS53A, located in Grendon

SOE agents liaised and worked with an estimated two to three million active resistance fighters in Europe alone [5]. A look at the range of recruits, all of whom were volunteers, would tell how desperate Churchill and the British government were. Recruits included women (including single mothers and divorcees), academics, Communists, former residents of Nazi-occupied countries, known homosexuals, prostitutes, and criminals - some of whom were recruited for their lock-picking ability among other 'talents'. SOE's ranks included the Oxford/Cambridge educated; escaped European aristocrats; and also members of the lower-classes of British society. Jewish paratroopers from Palestine who had earlier fled from Europe were also among the volunteers.

1.3 SOE Training

Some recruits were as young as nineteen and others were over the age of 50. Against the prevailing norms of class-conscious, homophobic and male-chauvinistic British society of the time, the SOE was a purpose-driven, meritocratic and truly 'equal opportunity' employer. An essential qualification for an agent was the ability to speak fluently the language of the area in which he or she would be inserted. Thorough familiarity with the local culture was also essential.

Training for many of the agents included firearms and explosives; deception and camouflage; map reading; parachute skills; and radio operation, including encryption and decryption. Twelve words per minute Morse was the minimum required receiving capability [6]. Noor Inayat Kahn, the first female radio operator to be inserted into occupied France, reputedly had the best "fist", sending Morse code at 18 words per minute (wpm) and capable of receiving at 22 wpm [7]. The Morse key used by the SOE agents was always a straight key – electronic keyers had not yet been invented! While the SIS (the Secret Intelligence Service [8]), did initially help the SOE by providing weapons and radios, the relationship between the SIS and SOE was not always friendly or cooperative. At times it was even hostile. Since the SOE did not have its own aircraft, it depended on the RAF for air transport of agents and for delivering provisions to its agents and resistance fighters in occupied territories. The Royal Navy only offered minimum support.

1.4 SOE Campaigns

The SOE turned out to be a very effective paramilitary organization even though many 'professionals' in the regular military establishment, as well as some in SIS thought the SOE something of a rag-tag band of amateurs engaged in "ungentlemanly" and at times "immoral" activities. However, according to both a German source and General Dwight Eisenhower, SOE's activities helped hasten WWII's end by half a year [9].

Many exploits of SOE agents have been revealed after having been kept officially secret for many decades after the war. There are others that will most likely remain undisclosed for some time to come, if ever. Many bridges, rail lines, telephone exchanges and landlines were destroyed, and assassinations were not unknown. For example, the Michelin tyre factory in France had provided tyres to the occupying German army - it was sabotaged. In Czechoslovakia in May 1942, SOE agents were instrumental in the assassination of Reinhard Heydrich, the chief architect of Hitler's 'Final Solution' to exterminate the Jews. Elsewhere, SOE orchestrated an attack on the Norsk Hydro plant manufacturing heavy water at Telemark, Norway; an operation that slowed the development of German nuclear bombs and was a collaborative work of SOE and local Norwegian resistance fighters. As well, SOE agents working with local French resistance fighters made major contributions to the successful outcome of Operation Overlord, better known as D-Day, by sabotaging German supply lines and their transport and communication systems [10]. In Asia, one of the most spectacular known works of SOE was blowing up the railway bridge being built in Burma by the Japanese Imperial Army using slave labour. This event was made famous by the film *The Bridge on the River Kwai*, with dramatic literary license being given to the actual events and location [11].

In addition to SOE and SOE-assisted local resistance activities, exiled governments from Nazi occupied countries were engaged in their own espionage, sabotage, and resistance activities. For example, the BCRA (Bureau Central de Renseignements d'Action) or Central Bureau for Intelligence and Operations, the French intelligence service under General Charles de Gaulle in London, inserted its own agents into occupied France. A resistance movement also arose in Norway immediately after Germany invaded the country in April 1940, as many Norwegians who escaped were trained in Britain and Canada and then dropped back into Norway. Likewise, the Belgian government in exile, first in Bordeaux, and later in London after the French surrender, called for the creation of organized resistance, and inserted British-trained resistance fighters into Belgium. Across Nazi-occupied west European countries generally, SOE agents helped with the training of local resistance fighters, supplying finance, the dropping of weapons and equipment (including radios), and providing other much needed supplies.

For SOE agents to carry out their missions, regular communication with headquarters in England was essential in order to receive instructions, report ongoing activities, and to request provisions. In the next chapter we will examine how radio was deployed in the field to accomplish this.

References

[1] The code name given to decryption of Axis messages. The Manhattan Project, which produced the world's first atomic bomb, is another closely kept major secret of WWII.

[2] Baker Street was where the office of the famous fictitious private detective, Sherlock Holmes was located.

[3] Judith L. Pearson, *The Wolves at the Door: The true history of America's greatest female spy*, The Lyons Press, Guilford, Connecticut, 2008, p.63.

[4] M. R. D. Foot, *SOE The Special Operations Executive 1940-46,* British Broadcasting Corporation, London, *1984,* p.62.

[5] Shrabani Basu, *Spy Princess: The Life of Noor Inayat Khan*, Omega Publications, New Lebanon, New York, 2007, p.55.

[6] Pearson p.71.

[7] Basu p.73.

[8] Also known as MI6, Military Intelligence 6 was the counterpart to the American OSS, the predecessor to the CIA. It was also called by another cover name, 'ISLD' (Inter Service Liaison Department), when referred to operation outside of the UK. Insiders were known to have called the organization "The Firm." Geoffrey Pidgeon personal correspondence, May 18, 2018.
[9] Rita Kramer, *Flames in the Field: The story of four SOE agents in occupied France*, Penguin Books, London, 1995, p.233.
[10] In the first days after the Normandy landings, 571 rail targets were sabotaged. Pearson p.201.
[11] The bridge that the SOE agent-assisted group blew up was actually over another river. Known under its general cover name "Force 136", the SOE's South East Asian operation was headquartered in Ceylon (now Sri Lanka). http://cambridgeforecast.org/blog2/2011/03/14/force-136-in-bridge-on-the-river-kwai/

2

How SOE Agents Operated Behind Enemy Lines

SOE AGENTS generally worked as teams of three in the field: a leader, a courier, and a radio operator. Some agents travelled into and out of Nazi-occupied areas through the borders of neutral countries such as Spain, Switzerland, or Sweden, using the neutral country's passport or other fake identification documents. Agents pretended to be journalists, salespeople, actors or members of other occupations. Some agents were embedded via submarines, fishing boats or feluccas. Sometimes insertions were accomplished with the help of the Royal Navy, which was often a reluctant partner of SOE campaigns [1].

2.1 The Drop

Agents were often dropped-off or picked up by a single-engined Westland Lysander, a STOL (short take-off and landing) aircraft capable of landing onto, and taking off from, a 300 to 400 metre-long strip, created on grass or on farmland in the countryside (Pic 2.1).

The Lockheed Hudson (**Pic 2.2**) was a longer-range and more powerful twin-engined

Pic 2.1 Westland Lysander airplane

Pic 2.2 Lockheed Hudson airplane

aircraft, requiring an airstrip twice the length of that needed by a Lysander, but was used when greater space was available.

Some were parachuted-in from even larger aircraft, such as the four-engined Halifax (**Pic 2.3**).

Hugh Verity was a pilot who flew both the Lysander and the Hudson, whose memoir contains a detailed account of how the flights were planned and executed [2]. The Lysanders and the Hudsons used on these missions flew only by moonlight, ie, only within a week before and after the full moon. Neither aircraft used radar or beacons to navigate. The pilot had to read a map and follow natural topographic markers such as rivers, lakes, and mountain peaks, as well as known landmarks such as church spires and other buildings whose lights may have been left on at night. Lewis M. Hodges, Verity's colleague in No 161 Squadron, flew the four-engined Halifax, which had been specially outfitted for parachute drops of agents and stores for the SOE. He wrote:

Pic 2.3 Halifax airplane

"Our lives were governed by the phases of the moon. We needed moonlight to map-read by; we needed moonlight to find our way to the dropping zones for parachuting, and to the small fields that served as landing grounds; and we needed moonlight to be able to see the ground early enough to make a safe landing." [3]

On arrival at the landing strip, a 'reception committee' would assemble. Three people formed an inverted L (or capital gamma Γ) at the landing site. One (A) placed himself at the bottom of the leg, which was about 100 metres long, the second (B) was located at the shoulder, and a third person (C) was at the end of the arm, about 50 metres away; each held a torch or a flashlight. Either the person at (A), or a fourth person, flashed a prearranged signal (a letter of the alphabet in Morse) to the pilot, to indicate that it was safe to land. The pilot then landed as close as possible to (A) with the torch light on his left, taxied up to (B), turned around between (B) and (C), taxied back to (A), and let his passengers disembark. The aircraft then took off.

These 'reception committees' helped not only arriving agents, but also collected supplies that were dropped, such as radios, weapons, explosives, local currencies, and fake documents (such as identity and ration cards), quickly gathering in the parachutes, and hiding or disposing of evidence of the arrival and the supply drop.

Pic 2.4 Brochart alias 'Charot', and Lartalias alias 'Tral', in Amiens, France. Parachuted-in during the night of April 8 to 9, 1944, in Neuvy-Pillous (Indre department) they transmitted with a Paraset from a garage located in the city of Amiens

The maneuver was so well practiced that the whole business could be accomplished in under three minutes, although it was an exceedingly risky operation. In the first year, fifty-five percent of SOE flights from England failed. Reasons for failure included adverse weather, failure to find the landing zone, failure of the reception committee to show up, or of course, the arrival may have been compromised by the presence of enemy troops or agents. The situation improved over time, but even by 1944, only two-thirds of flights succeeded [4].

In most cases, after arrival, an SOE team would liaise, and coordinate their sabotage and espionage activities with local resistance fighters (**Pic 2.4**). Each grouping of SOE agents and local resistance fighters was called a 'circuit' (réseau) and each circuit, as well as each agent, was given a code name. However, these code names were changed from time to time when circuits or agents were compromised, or when there was a suspicion that the Germans had found out who they were. Examples of circuit names include Prosper, Cinema, and Heckler, while Diane was a code name for one of the most accomplished American agents, Virginia Hall, who worked for SOE and later for the OSS. Madeleine was the code name for SOE radio operator Noor Inayat Khan, and Gilbert for another radio operator, agent Henri Déricourt.

2.2 Communicating with Agents in the Field

Radio, (then called 'wireless' [5]), was the principal method of communication between SOE headquarters and circuits in the field. All SOE wireless communication was sent by Morse, with the text encrypted before transmission from the UK [6]. Powerful transmitters, with large antennas installed high up, ensured not only stations with small aerials in the field and behind enemy lines could receive SOE signals, but that receiving stations in the UK could copy weak signals sent from the field.

Communications to and from field agents involved receiving orders from SOE Headquarters in England; reporting how the organization and execution of sabotage plans by local resistance fighters were proceeding; relaying information about compromised or thwarted operations; sending requests; and confirming arrangements for the dropping of provisions and the supply of weapons. Messages from the field reporting infiltrated circuits, or captured agents, were not infrequent. Specific information regarding drop-off and pick-up points in the field, and the confirmation code required by the pilots, was also exchanged using these encrypted communications to ensure that arrivals and departures were as secure as possible.

Another technique for delivering important directives and messages from headquarters was to hide them in AM broadcasts. Such messages were regularly broadcast in BBC programs that were being beamed towards the Nazi-occupied countries. To France alone the BBC broadcast eight or nine programs daily, with the main program of the day being transmitted at 7-30pm every evening [7]. Hence, in the middle of an entertainment program on the BBC French language service, the announcer might say, "Jasmine is playing her flute". This was a coded message, telling field agents to prepare for arriving agents [8]. The reply to this message was transmitted back to London in a coded Morse message. That particular message was one sent to agent Noor Inayat Kahn, and is, of course, a form of cross-mode (AM-CW) QSO in Amateur Radio parlance.

As modern Radio Amateurs know, many successful contacts can be made between a very low powered (QRP) station, and a station operating at much greater power (QRO) which has more sophisticated antenna equipment. The high-powered station does the heavy lifting, using a high-powered transmitter, an efficient antenna and a more sophisticated and sensitive receiver than that of the lower-power operator with a relatively small antenna in the field. Radio communication between an SOE agent in the field, and a base station in England, was always such QRP to QRO operation. Radio operators in the field carried portable transceivers with an output power of only five to twenty watts, whereas SOE had two home QRO transmitting stations: Grendon Underwood, and Poudon, located between Bletchley and Thame. UK-based stations normally used 250 watt transmitters, but they also had a 15kW transmitter for use when propagation conditions were not favourable [9]. SOE had the benefit of many receiving sites with large antennas, manned (or 'womanned' would be a more precise term as the majority of W/T operators at these base stations were women) around the clock. Furthermore, a large team of women worked in encryption and decryption at Bletchley Park, and nearby at what are called Bletchley Park's outstations.

In the field, a far greater portion of the radio operator's time was spent encoding outgoing messages prior to transmission, and in decoding received messages, than in the actual transmission and receipt of messages over the air. Messages were initially enciphered by SOE using what was called a Playfair code [10]. In its basic form, it was relatively easy for the Germans to break. However, when a young genius code-master named Leo Marks joined SOE and headed the encryption team, the situation improved, but, there followed a constant cycle of making the coded messages increasingly difficult to break, and the Germans' increasing ability to break the encrypted messages. In 1942, Playfair was abandoned as too risky, and in its place a double-transposition ciphering system was introduced, although even that was not unbreakable. Under Marks' leadership, SOE finally settled on the 'one-time pad' for encryption. With this approach, an agent and the home station each held the only copies of pads of pages comprising random figures and letters used to encrypt and decipher each message. After being used once, a given page of the pad was destroyed. When used properly, the one-time pad was regarded as the most highly secure approach.

2.3 The Dangers

All agents in the field faced enormous danger. They were made fully aware of the risks involved when they signed on. If caught by the Gestapo (the Nazi State Secret Police), the Abwehr (the German military intelligence organization), or a local police force collaborating with the occupying Germans, agents would face certain torture to obtain information, or for revenge, deportation to concentration camps in Germany, and finally execution. Collaboration by official bodies in occupied zones, such as the Milice in France, and the Nasjonal Samling in Norway, was well known and understood. The treatment of captured agents was particularly brutal and cruel because they were not in uniform, not considered regular combatants, and therefore not entitled to the protections of the Geneva Convention. The SOE inserted a total of 480 agents into France, whose activities were the most numerous in all of Europe. Twenty-five percent never came home. Forty of those agents were women [11].

Among agents, radio operators faced the highest level of danger. Referred to as 'pianists', their survival time in the field was estimated to be six months [12]. They were often caught with the most incriminating physical evidence of their job - a radio. Unfortunately, a wire antenna, perhaps 70 feet long, needed to be hung, preferably outside their transmitting location, and could often be spotted by neighbours who might report it to the Germans for a reward (see Noor Khan's story Chapter 2.5).

No VFO (Variable Frequency Oscillator) - equipped radios were used; transmitting frequencies were generated from a choice of crystals that could be plugged in for stable signal generation at the assigned frequency. In the event of discovery, it would be difficult to concoct an innocent explanation as to what the crystals were really for.

Every time an agent transmitted, his/her radio signals were essentially announcing the location of the transmitter to the German direction finding (D/F) networks. Moreover, many transceivers' regenerative or super-heterodyne circuits generated low-level RF signals even when receiving. The German D/F networks were very sophisticated and efficient. They kept a 24/7 watch on every radio frequency and it took them only twenty to thirty minutes for their direction-finding teams in the field to get within a few yards of an operator who stayed on the air too long. For example, on Avenue Foch in Paris, the location of the Gestapo headquarters, thirty clerks kept a continuous watch on cathode-ray tubes monitoring the radio spectrum. Every time any new signal appeared, its frequency was distributed to the direction finding/listening posts in Brest, Augsburg and Nuremberg in order to triangulate the location of the signal source to within a few miles. Within fifteen minutes, vans with D/F antennas were closing in on the area indicated [13]. Once within a few hundred yards of the signal source, German agents wearing overcoats with antennae inside the sleeves, and with an RF signal strength meter in hand, would come in for the 'kill' (**Pic 2.5**). Another method used to locate a transmitter site, was to switch-off mains power section-by-section in a town, or floor-by-floor in a tall apartment building, until the transmission suddenly ceased as the mains current feeding the radio was cut-off.

Pic 2.5 Gürtelpeiler (portable radio direction finder)

At first, the clandestine radio operators had orders from London not to transmit for more than fifteen minutes from the same location [14]. But by the winter of 1943-44 SOE had ordered that no radio transmission was to last more than five minutes [15]. The danger was not only to the radio operators themselves, but also to local residents who provided their houses or farm buildings for such radio operations, or who helped hide agents' radio equipment. Locals friendly to the Resistance were tasked to serve as lookouts for Germans who might be searching for SOE radio operators. Many such locals were caught by the Germans, or turned over to them by other locals who were collaborating. When discovered, they were deported to the concentration camps.

Most crucially of course, the radio operator's key objective was to carry on without being caught by the enemy, whether military, local police or collaborating residents. Finding safe locations to operate, hanging antennas, and changing the transmitting/receiving locations were constant and essential chores.

Transporting the radio and its power supply was not an easy task, despite the commonly used Paraset transceiver (the Whaddon Mark VII) being relatively light and portable inside a small suitcase or in a shopping bag. This is in contrast to the higher powered transceiver (the Whaddon Type 3 Mark II - also known as B2) which was introduced in 1942, whose transmitter ran about 20 watts output, but weighed about 35 lbs, and required a much larger container than the Paraset. It was usually carried in a suitcase that was two feet in length, so the agent carrying such equipment needed strong nerves and quick thinking while transporting his or her gear by car or bicycle, or on public transport, as suitcases were regularly inspected at the many check points [16].

One of the most memorable stories of a narrow escape was related by M.R.D. Foot, regarding an agent whose code name was Felix [17]:

"..., a Jew of Alsatian-Polish origins who was assistant wireless operator to the young 'Alphonse', a British agent in southern France. He, 'Alphonse', and 'Emanuel' the wireless operator [a quiet Canadian] all got out of the same train at Toulouse; 'Felix', carrying the transmitter in its readily recognizable suitcase, went up to the barrier first. Two French policemen were conducting a cursory check on identity papers. Behind them, two uniformed SS men were sending everyone with a case or big package to the *consignee*, where more SS men were making a methodical luggage search. 'Felix' took in the scene; ignored the French police; held his suitcase high; and called in authoritative German, 'Get me a car at once, I have a captured set.' He was driven away in a German-requisitioned car; had it pull up in a back street; killed the driver, and reported to 'Alphonse' with the set for orders."

2.4 Capture and Deception

A captured SOE radio operator was often forced to continue sending and receiving false messages to and from England. Or sometimes, a confiscated radio may have been operated by a German operator who would attempt to imitate the operating style of the captured SOE agent.

SOE radio operators were trained to include security checks hidden in their messages to signal that they were sending messages under duress. This involved sending deliberate errors at a designated position in a message, such as every sixteenth letter in the text, or in never sending a sentence which contained a word with a specific number of letters. When Noor Inayat Khan sent a message that contained an 18-letter word, Leo Marks instantly knew she had been compromised, although others in the SOE hierarchy did not believe him at the time.

On one occasion, to ascertain whether or not an SOE radio was being operated by a German, Marks had an SOE home operator slip the Morse code 'HH' in at the end of a transmission. Now it had become an instinctive habit of German operators to put 'HH', standing for 'Heil Hitler', at the end of their internal exchanges [18], so when the 'SOE' operator in the Netherlands, without hesitation, responded by sending 'HH' back, it was clear that the radio was being operated by a German.

One of the most disastrous failures of SOE radio communication occurred in the Netherlands, where over 50 SOE agents were captured, tortured, and executed. The Abwehr called their operation *Das Englandspiel* (The England Game), it was also called *Unternehmen Nordpol* (Operation North Pole). Unbeknownst to SOE HQ, the Germans had captured agents blindly dropped by SOE into the Netherlands in 1942, and used the SOE radio operators as well as the agents' codes to fool the British into sending more agents and supplies. As apprehended agents sent coded messages without sending the required security checks, Leo Marks realized that the agents had been compromised. However, others in the SOE leadership were not convinced, so the SOE operation in the Netherlands was not shut down until April, 1944. When Herman Giskes, head of the *Englandspiel* in the Netherlands, realized that the German penetration had been uncovered by the British, he sent a cynical message in clear text [19]:

"To [SOE section chiefs] Messrs Blunt, Bingham and Succs Ltd., London. In the last time you are trying to make business in Netherlands without our assistance STOP We think this rather unfair in view of our long and successful co-operation as your sole agents STOP But never mind whenever you will come to pay a visit to the continent you may be assured that you will be received with the same care and result as all those who you sent us before STOP so long"

There has been a persistent theory, even to this day, that the SOE leadership knowingly continued to pretend that the Dutch operation had not been compromised and kept sending agents and stores for strategic reasons, most notably, to make the Germans believe that the Allies' invasion of Europe would take place in the Netherlands and not Normandy. This conjecture has never been substantiated. It is likely that such deliberate disbelief, or the pretension, occurred as a result of the inter-agency enmity between the SOE and the SIS/MI6. For internal political reasons, no agency admitted major errors or any betrayal of agents.

In his post-war memoir, Herman Giskes wrote and described how the Germans captured SOE agents and kept them sending messages to England under duress until 1944. Reading Leo Marks's *Between Silk and Cyanide: A Codemaker's War, 1941-1945,* and Herman Giske's memoir *London Calling North Pole* (translated from the original German edition *London ruft Nordpol : das erfolgreiche Funkspiel der deutschen militärischen Abwehr*) in juxtaposition is an interesting exercise. The opposing perspectives aid in understanding how SOE's coded messages were developed, handled, and broken [20].

In tribute, Brigadier Colin Gubbins, who led SOE from 1943 as its Executive Director, said that the work of SOE radio operators was "the most valuable link in the whole of our chain of operations. Without these links, we would have been groping in the dark" [21]. Likewise, Noreen Riols, who worked at SOE headquarters and knew most of the agents embedded in France, said "The radio operator, being indispensable, was possibly the most important of the three agents [who comprised each SOE team] since it was through him that contact with London was made to request supplies or replacements, instructions and also calls for help..... [H]e was the one member of the team the most exposed to risk. An organizer or a courier could 'lie low' if they suspected that they were being watched. A radio operator could not." [22].

2.5 Three SOE Radio Operators in Action

Many books have been written about SOE and the agents' work in the field, but records documenting specific day-to-day, minute-by-minute, activities of a radio operators' work are scarce, which is true to the nature of their job. The existence, let alone specific activities of SOE and its agents, were not only kept secret during the war, but for many decades afterwards. There are only two books that the author has come across which document in relative detail, how SOE radio operators worked and how their radios were deployed on a daily basis:

Noor Inayat Khan

In *Spy Princess: The Life of Noor Inayat Khan*, and other accounts memorialized by Khan's personal acquaintances [23], there is a well-documented record of how she was trained as a radio operator, and how she used her radio (a Type 3 Mk2, also known as the B2) in Nazi-occupied Paris. Hers was the remarkable story of an unlikely young French-Indian woman who volunteered to be an agent. She worked as a very effective, but lone radio operator, from the Paris area for a few crucial months in 1943. Khan was only nineteen years old and a most unusual candidate when she volunteered to be an agent. She was a musically and artistically talented, dreamy daughter of a Sufi leader. Hers was an aristocratic Indian lineage, and her mother was American-born, hence the Princess moniker. During her training she mastered Morse code very quickly (18 wpm sending, 22 wpm receiving), which was a remarkable achievement after only a few

Pic 2.6 Noor Inayat Khan

months of training. This is a level of proficiency that many Amateur Radio operators only reached after several years of diligent practice, in an era when Morse code was still required for an Amateur license; indeed many Amateurs have never reached her level of proficiency. She had been inserted into a Paris suburb by Lysander on the night of June 16/17, 1943. By the time she was betrayed and arrested by the Gestapo on October 13 1943, she had sent hundreds of messages, skilfully eluding her Germans pursuers during that brief period. She moved from one location to another every few days, or few hours, carrying the heavy radio and power supply, and often had to climb to a roof to set-up the antenna wire.

Unfortunately, Kahn was caught by a German soldier hanging a wire antenna on a tree; she cleverly explained that she was hanging-out a wire to dry clothes. Helped by her charming and demure personality, she successfully persuaded the German, and on that occasion, had a narrow escape, only to later be betrayed by a neighbour and caught just two days before she was due to be picked up from a secret air field and returned to England. Leo Marks immediately knew she had been caught when she transmitted a message containing an 18-letter word, a prearranged security check to be sent only when needed to indicate she had been caught [24]. Soon after that, the receiving operator noticed that Khan's key touch had changed and became convinced that the Germans were operating her radio set [25]. All agent radio operators in the field had been "fingerprinted" before they were sent behind the lines, ie the characteristics of their Morse sending "fist" were recorded in case the enemy tried this type of imitation and deception. [26].

Khan was sent to the concentration camp at Dachau, Germany, and was tortured and executed along with two other female SOE agents on Sept. 13, 1944. She was posthumously awarded the Croix de Guerre and the George Cross, the highest honour for heroism in France and Britain respectively. See the film *Enemy of the Reich.*

Pic 2.7 Oluf Reed Olsen operating a Paraset in the field.

Oluf Reed Olsen

Two Eggs on my Plate, is an autobiography by Oluf Reed Olsen a Norwegian resistance fighter who escaped to Britain from Norway soon after the Nazis invaded in 1940 [27]. He was trained by SIS/MI6, SOE, and the exiled Norwegian military organizations in Britain and Canada, and parachuted back on April 20, 1943. His reporting of German ship and troop movements helped sink many German boats and sabotage German transport. He had many narrow escapes (described in detail in the book), but was able to leave Norway via Sweden just ahead of Nazi pursuers, and lived into his 80s.

Olsen's radio is on a permanent display at the Imperial War Museum in London today (see cover photo). The display includes not only his Paraset transceiver, but also some code tables, including a one-time use cipher pad, and transcripts of some of the messages he sent. The description incorrectly identifies the radio as Mk II but it is clearly a Mk VII, the Paraset.

Virginia Hall

One of the most celebrated and successful SOE/OSS agents, Virginia Hall played an important leadership role. She organized a large local resistance group, and in July 1944 was waiting for a major drop of weapons and French currency. She waited several days for personal messages that were a regular feature of the BBC broadcast [28]. Some of these broadcast messages were genuine, but some were completely nonsensical phrases designed to confound the enemy. On July 21, 1944, she finally heard "Les marguerites fleuriront ce soir" (the daisies will bloom tonight) followed by "Je dis trois fois" (I say three times) and told her resistance group that the drop would be made by three planes that night.

On April 4, 1944, she radioed to London that her agent had successfully found safe houses for Allied airmen in Paris. London replied asking the number

Pic 2-8 Virginia Hall operating a B2 spy set.

they could accommodate. She replied on her next "sked" (scheduled contact) and gave the secret pass phrases to use when approaching each house. London replied they weren't able to copy her. They suggested that she extend the antenna, which she did, but to no avail [29].

Communication during WWII was assisted by a period of relatively high sunspot activity which permitted a low-power radio signal to propagate more effectively than it might have done during other phases of the sunspot cycle. Nonetheless, radio signals didn't always propagate as well as expected as the above example shows.

The next chapter examines the technical details of the radios used by SOE agents and resistance fighters.

References

[1] Noreen Riols, *The Secret Ministry of Ag. & Fish: My life in Churchill's school for spies*, Pan Books, London, 2013, pp.28, 29.
[2] Hugh Verity, *We Landed By Moonlight: The secret RAF landings in France 1940-1944*, Crécy Publishing Limited, Manchester, 2000 revised edition.
[3] ibid p7.
[4] M. R. D. Foot, *SOE The Special Operations Executive 1940-46,* British Broadcasting Corporation, London, *1984.* pp98, 99.
[5] In British official documents, radio telegraphy and radio operators were abbreviated as W/T, standing for wireless telegraphy, and wireless telegraphist.
[6] The words code (encode, decode) and cipher (encipher, decipher) are often

used interchangeably in common parlance. But there is a difference between them: a code is a method of encrypting on the level of meaning, ie, a code represents a word or a phrase. A cipher operates on a level of a letter, ie, a letter may be expressed by another letter or a number. A message may first be encoded and then enciphered. A code cannot be decoded without a code-book. One of the never-broken codes used during WWII is that of the Navajo code-talkers who served in the Pacific theatre of war. See for example, Sally McClaim, *Navajo Weapon: The Navajo Code Talker*, Rio Nuevo Publishers, Tucson, Arizona, 2002.

[7] Riols, pp.50, 51.

[8] Shrabani Basu, *Spy Princess: The Life of Noor Inayat Khan*, Omega Publications, New Lebanon, New York, 2007, p.103.

[9] M.R.D. Foot, *SOE in France*, p.110. The author's own personal QRP QSO experience includes a QSO between a portable setup on a high-rise rooftop in Manhattan, New York, and Sendai, Japan, about 6500 miles apart. The author was transmitting 2.5 watts SSB from a Yaesu FT817 with a short portable vertical antenna and the Japanese station was running 1 kilowatt with a five-element Yagi about 20 meters high.

[10] *ibid* pp. 122-124. See Appendix 1 for technical details of how Playfair and later coding systems worked.

[11] Basu, p.105.

[12] Judith L. Pearson, *The Wolves at the Door: The true history of America's greatest female spy*, The Lyons Press, Guilford, Connecticut, 2008, p. 245.

[13] Foot, *SOE,* p.106.

[14] Riols, p. 33

[15] Foot, *SOE,* p.106.

[16] The specifications for these radios are discussed in the next chapter.

[17] Foot, p.111.

[18] Leo Marks, *Between Silk and Cyanide: A codemaker's war 1941-1945,* The Free Press, New York, 1998, p.348.

[19] ibid p.499.

[20] H. J. Giskes, *London Calling North Pole: The true revelations of a German Spy*, Echo Point Books & Media, Brattleboro, Vermont, 1953.

[21] Foot, *SOE in France*, p.95.

[22] Riols, p.32.

[23] Various aspects of Khan's activities and experiences were recalled in a number of publications by those who personally knew her, including Leo Marks', *Between Silk and Cyanide,* Noreen Riol's *The Secret Ministry of Ag. & Fish*, and Hugh Verity's *We Landed by Moonlight.*

[24] Leo Marks, *Between Silk and Cyanide: A codemaker's war 1941-1945,* The Free Press, New York, 1998, *p.399.*

[25] *ibid* p.411.

[26] *ibid* p.601.

[27] Oluf Reed Olsen, *Two Eggs On My Plate,* Rand McNally & Company, Chicago, 1952. Translated by F. H. Lyon from the Norwegian titled, *Vi kommer igjen* (1952).
[28] Pearson, p.211.
[29] Pearson, p.186.

3

Equipment used by SOE Agents in the Field

THE RADIO used by SOE agents behind enemy lines and local resistance fighters, required a number of features not usually expected of ordinary fixed-location-use radios. These features nonetheless are essential for the agent's task:

3.1 Summary of Requirements

1. Transmitter power output: it needed to produce sufficient CW power, and its signal needed to be capable of reaching England from the European continent, up to 300 to 500 miles distant.

2. Receiving capability: it needed to receive both AM (BBC broadcasts mainly) and CW signals transmitted from England.

3. Portability: it needed to be light, and small enough to transport in a small suitcase, rucksack, or travel bag, by train, bus, or bicycle.

4. Operability in the field: it needed to be capable of operating in the "flat" position on the ground. The usual vertical front panel arrangement of a fixed-station radio would not be as practical.

5. Disguise: the radio and its accessories needed to be small and easy enough to hide in homes, in public conveyances and outdoors. The case or housing of the radio needed to be disguised as an ordinary item for travelling or household use.

6. Durability: it needed to be sturdy enough to withstand para-drops and constant physical movement.

7. Power supply: it needed to be operable from AC mains, battery, or from a hand-cranked or bicycle generator.

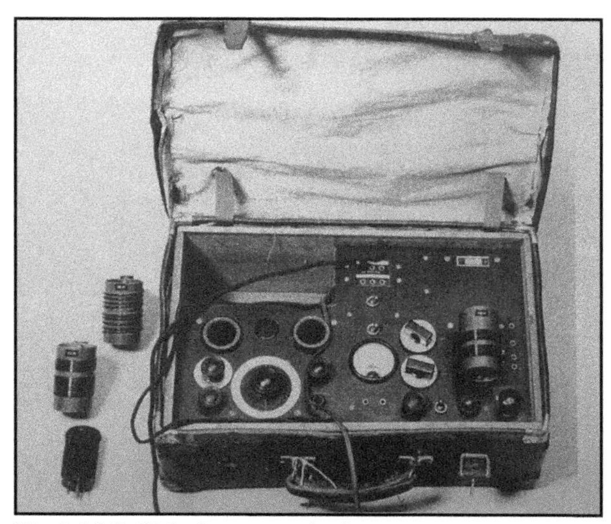

Pic 3.1 Mk V (suitcase version)

Pic 3.2 Mk V (Separate receiver & transmitter version)

8. Repair and maintenance: it needed to be easy to repair and maintain and utilize only easily obtainable parts.

3.2 The First SOE Radio - The Whaddon Mk V

While the main focus of this book is on the Paraset, it ought to be mentioned that there were other clandestine radios used by SOE agents and local resistance fighters in the field that met the above requirements. SOE's first radios were supplied by the SIS, because the SOE in its early days was not capable of manufacturing radios (or weapons, and other materials for that matter).

The very first radio offering was the Mark V two-valve CW transmitter, paired with a three-valve receiver [1] (**Pic 3.1 & Pic 3.2**).

It could be operated either from AC mains, or from a battery with a vibrator (an early DC-to-DC-converter). The units were housed in plywood boxes, and weighed 15 kg (33 lbs) altogether. This transmitter-receiver set was not only heavy, but cumbersome to carry and use. It had the following specification:

<u>Receiver</u>

Circuit features ('line-up'): RF amplifier + Regenerative detector + AF (AM, CW) + AF amplifier

Frequency Coverage: 3.7 - 7.5 MHz and 7 - 16 MHz using two sets of plug-in coils

AF Output: High-impedance headphones

Valves: 6SK7 (x3)

Schematic: see **Fig 3.1**

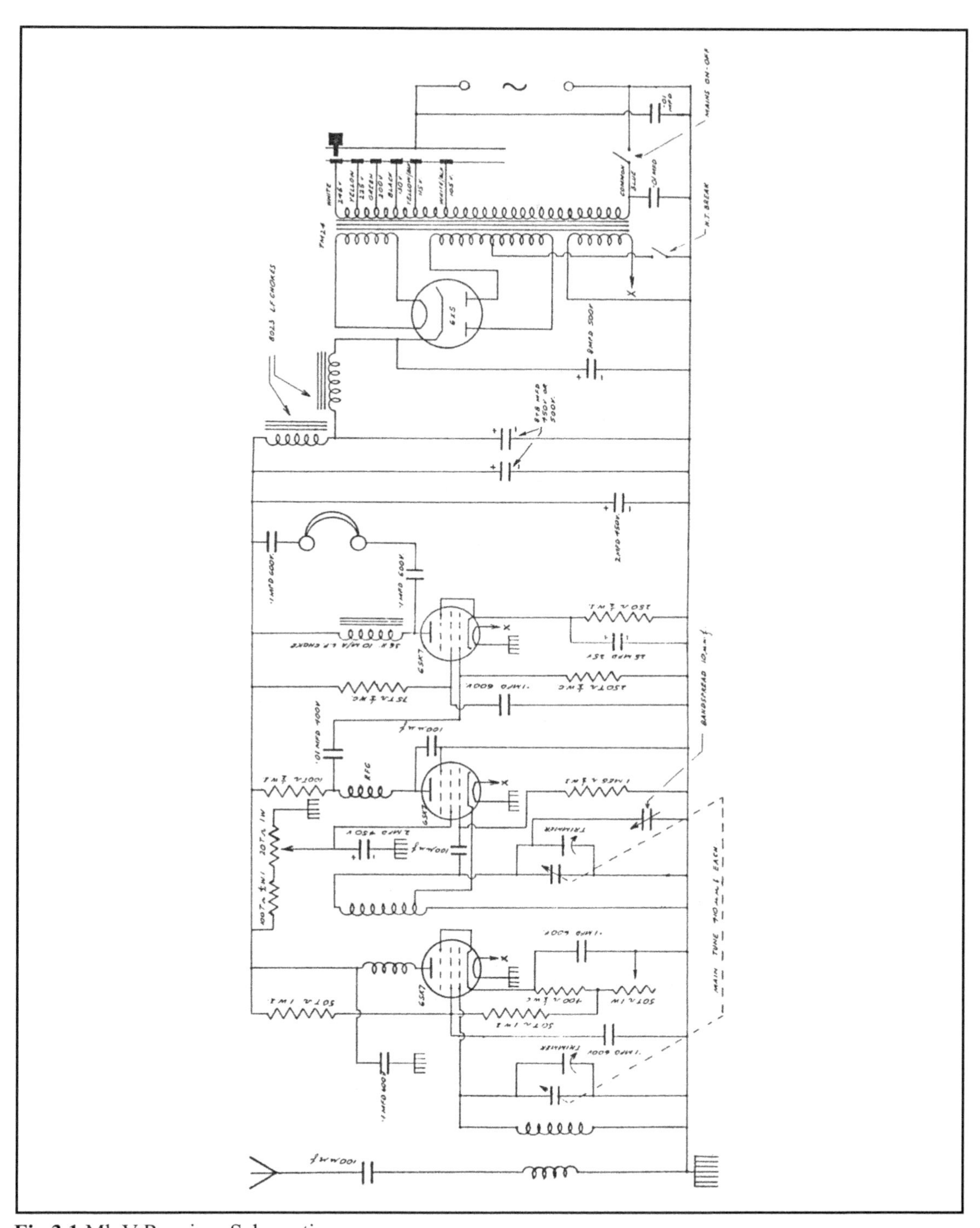

Fig 3.1 Mk V Receiver Schematic

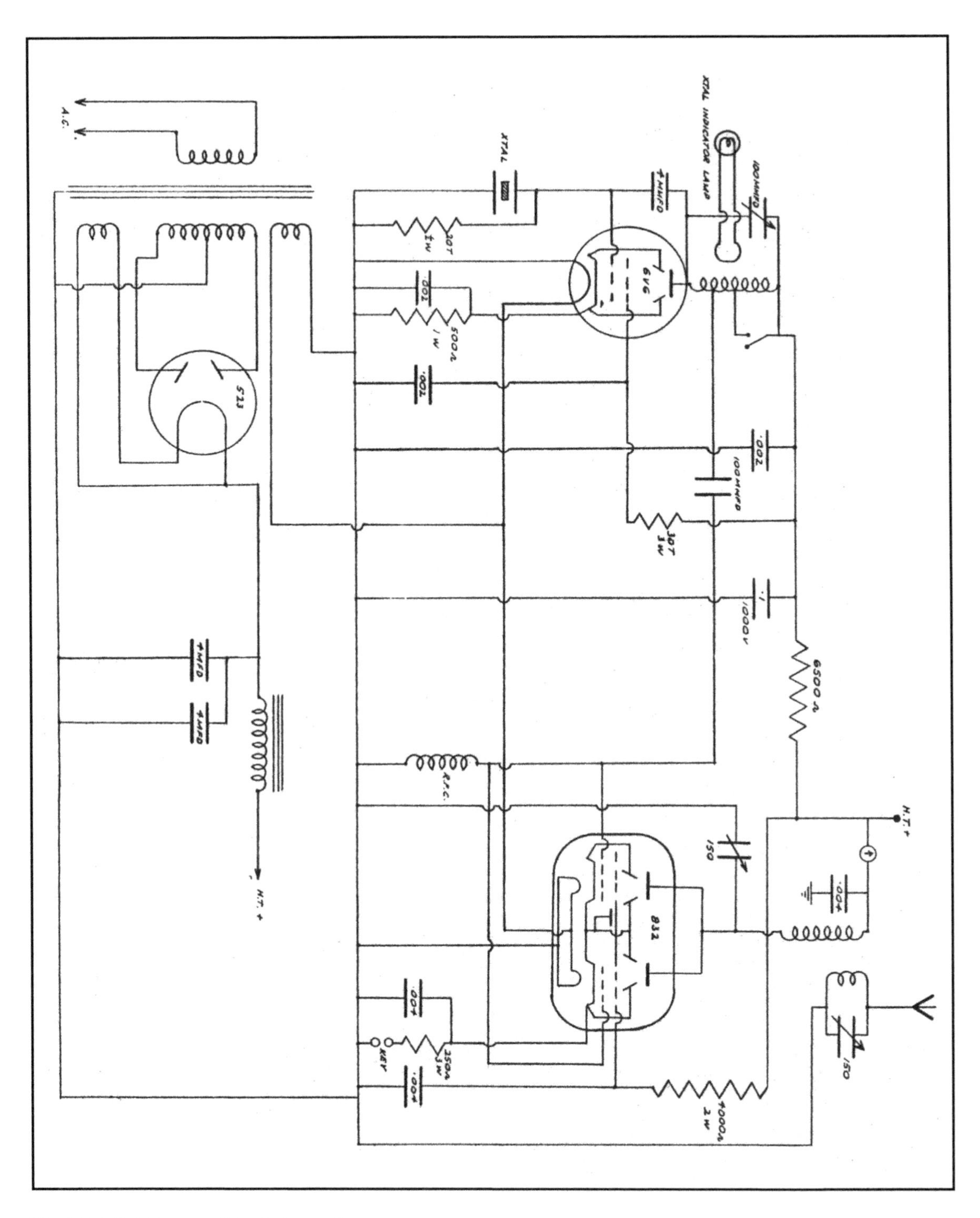

Fig 3.2 Mk V Transmitter Schematic

Transmitter

Circuit features: Crystal oscillator, RF power amplifier (CW only)

Frequency Coverage: 2.9 - 18 MHz, using three plug-in coils

RF Power Output: 20 - 25W

Valves: 6V6 and 832

Schematic: see **Fig 3.2**

The RF power amplifier could be used as a doubler for 9 - 16MHz. The transmitter has a built-in Morse key but provision is made for connecting an external key.

Power Supply

AC mains, 100 - 250V; rectifier 5T4 or 5Z3 (depending on version)

General

Size (cm) and Weight (kg): Complete set in suitcase:

Height	Length	Width	Weight
18	27.5	43	15

Antenna: Wire, 30m long, and grounding lead

Accessories: Crystals, antenna wire, grounding lead, coils, spares

3.3 The Next Generation SOE Radio – the Mk VII 'Paraset'

In 1940 SIS next developed a smaller, lighter, and much more manageable transceiver [2]. Designated the Whaddon Mark VII, it came to be known as the Paraset, and was produced throughout WWII. It was the radio most widely used by local resistance groups and SOE agents in the Nazi-occupied countries and also by agents in the Japanese-occupied areas of Asia [3]. Its characteristics were as follows:

Receiver

Circuit features: Regenerative detector + AF amplifier (AM, CW)

Frequency Coverage: 3.0 - 7.6MHz in a single range, with a separate fine tuning dial fitted to the front panel

AF Output: High-impedance headphones

Valves: 6SK7 (x2) metal version

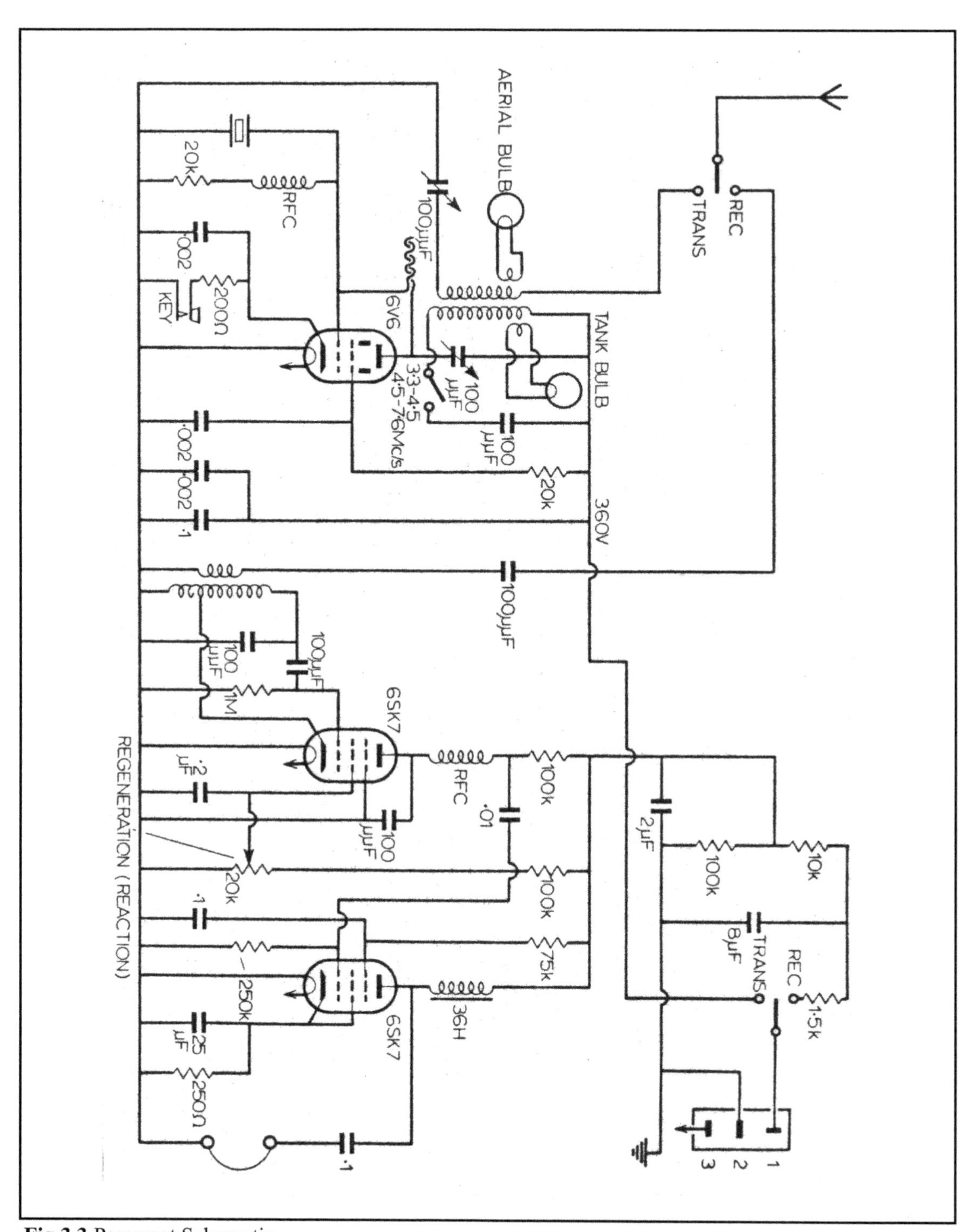

Fig 3.3 Pararaset Schematic

Transmitter

Circuit features: Crystal oscillator/RF power amplifier (CW only)

Frequency Coverage: 3.3 - 7.6MHz in two ranges:

Range 1: 3.3 - 4.5MHz

Range 2: 4.5 - 7.6MHz

RF Power Output: 4 - 5W

Valves: 6V6 metal version

Internal Morse key; provision for an external key in some versions

Power Supply

AC mains, 100 - 250V; rectifier 6X5; 6V battery vibrator power unit

General

Size (cm) and Weight (kg):

Item	Height	Length	Width	Weight
Transceiver:	11	14	22.5	2.3
AC Power Supply Unit:	11	11	14	2.7
DC Vibrator Supply Unit	11	11.5	14	2.9
Antenna:	Wire antenna and grounding lead			
Accessories:	Crystals, antenna wire, grounding lead, spares			

Overall circuit diagram: see schematic, **Fig 3.3**:

Two versions of the Paraset were made. The earlier version was housed in a wooden box (**Pic 3.3)** while the later version was built inside a steel box; nicknamed "the cash box" (**Pic 3.4**).

The latter came about to minimize RF leakage (unintended radio frequency emission) flowing from the regenerative circuit in the receiver, to outside the box. This was done in order to reduce the possibility of detection by German direction finders [4]. Both versions were typically carried in a small suitcase.

Pic 3.3 An early Paraset in wooden case

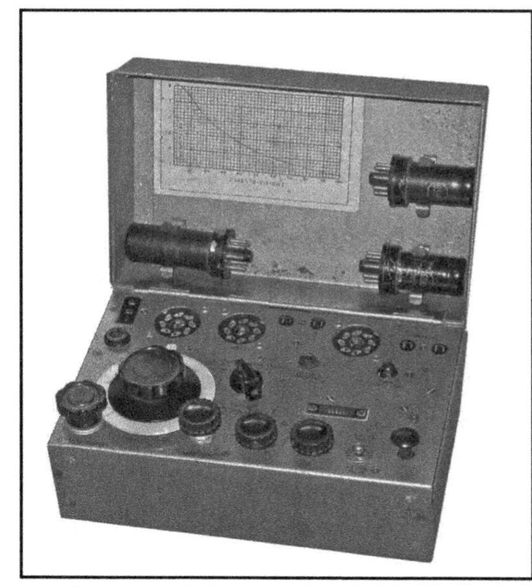

Pic 3.4 A later "Cash Box" Paraset in metal case

The SOE issued two types of quartz crystal for their spy radios: the Bliley type, with a round case, and a square-case crystal resembling the FT-241 with a footprint the same as the larger Bliley type (**Pic 3.5**).

The valves, metal-case versions of the 6SK7 and the 6V6, were American-made.

Though the straight CW (Morse) key is built into the front panel of the Paraset, some sets came with a provision for attaching an external straight key. SOE agents never used a mechanical speed key like the Vibroplex, perhaps because it was too delicate to transport and to maintain alignment in the field.

One notable "feature" of the Paraset is the absence of some of the usual features found in most modern transceivers. It lacks a volume control (you can adjust the audio level by changing the position of the headphones, or to some extent, by adjusting the regeneration/reaction controls). It does not use any meters; instead, antenna tuning and loading is indicated by two then-common 6.3v bayonet-style dial light bulbs. There are no relays employed to switch between transmission and reception; switching is done manually via a two-position rotary switch.

Pic 3.5 Quartz crystals for spy radios

Another missing feature is that of a "spotting" capability, ie, there is no way to spot on the receiving dial the exact frequency on which one is transmitting. In modern Amateur Radio HF communication, one normally transmits and receives on exactly the same frequency (co-channel working), but in clandestine communication of WWII, the transmitting and receiving frequencies were always different. In other words, the communication was always on a split-frequency basis.

Further views of the Paraset and its alternative power supply are shown in **Pics 3.6 to 3.8**.

The Paraset is truly a minimalist design; it is hard to think of any component that would be considered inessential. There are no 'bells and whistles' in this radio, yet, it could deliver everything needed to meet the demands of the time.

3.4 The Paraset in Use

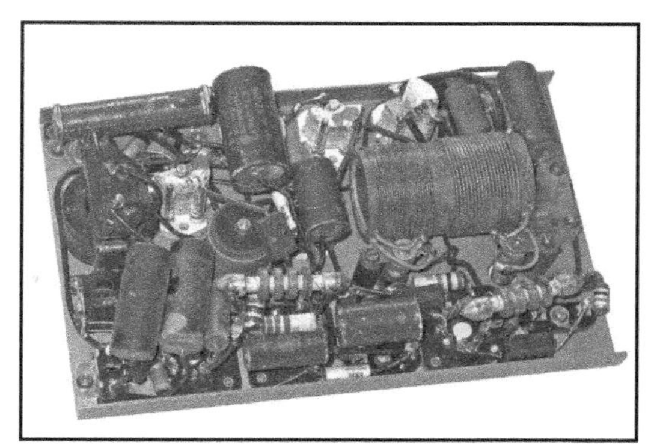
Pic 3.6 Underside view of Paraset

There is one vivid record of how the Paraset performed in the field, written by a BCRA agent Maurice de Cheveigné. Parachuted into Lyon, on May 31, 1942. He wrote:

"The main qualities of the Paraset transmitter-receiver that I use are that it is both small and light. The receiver, of the reaction (regenerative) type, is quite sensitive and can receive weak signals, but it is a bit temperamental to set and it varies with the fluctuations of the voltage coming from the mains as well as with the proximity of the operator's hand. It can hardly be considered as selective: a powerful station, close to the wavelength of your correspondence, will burst your eardrums and makes it difficult to read Morse. The transmitter has a low power: 4 watts. On short-wave, it's not an issue, especially as my correspondent in England has sophisticated receivers and the huge antennas of the "Home Station", STS 53A, located in Grendon.

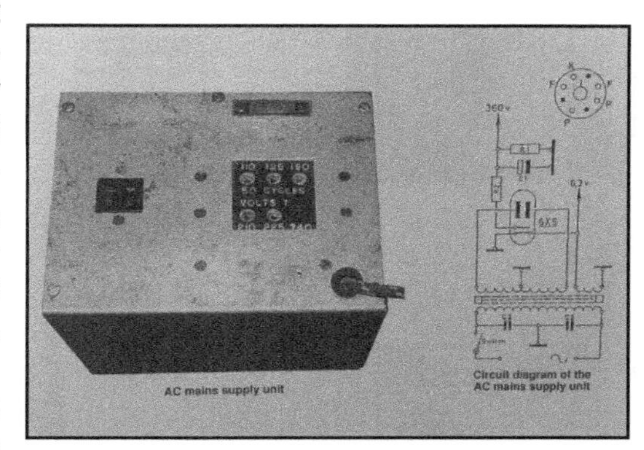

Pic 3.7 AC power supply

Ritually, I was also listening to the BBC on shortwave, to hear despite the intense jamming effort of the Germans, some news about the war. Marvellous short-wave helping the weak to get the upper hand on the master race!" [5].

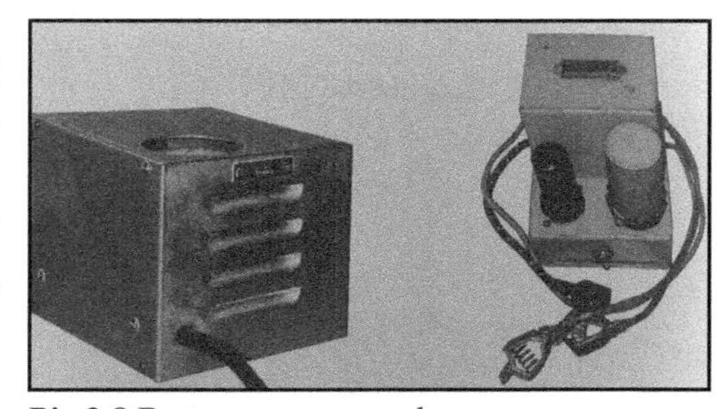
Pic 3.8 Battery power supply

The Paraset replicas the author has built behave in exactly the same manner as de Cheveigné describes. One other thing about the Paraset's transmitter is that it "chirps"; ie, there is a notable audio frequency/tone shift during the sending of a dash. This is typical of a single-valve oscillator/final valve transmitter whose power supply voltage is not regulated, as is the case with the Paraset. The Paraset is not the easiest radio to operate, but it did the job. As you will see later, a number of Radio Amateurs, including the author, have used replicas, built to the original specifications, on the 40m and 80m Amateur bands, and have regularly had successful QSOs over distances of several hundred-

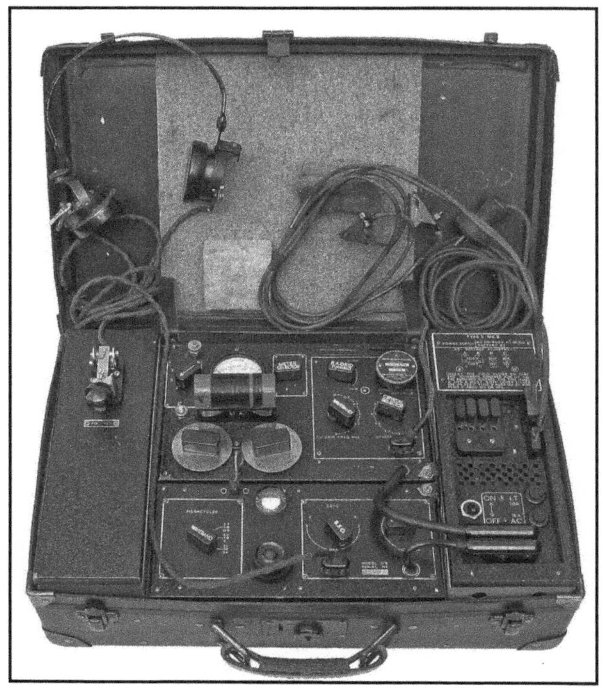
Pic 3.9 A Type 3 Mk II (B2) SOE transceiver

miles. In fact, in a few instances, the Paraset signal crossed the entire Atlantic from the US to the European continent and vice versa. A Canada to South Africa contact by a Canadian Amateur using a Paraset is also recorded.

3.5 Beyond the Paraset – The B2

On June 1, 1942, SOE's radio organization was completely separated from SIS, and by October 1942 SOE developed and added the Mark II transceiver. The subsequently revised version was the Type 3 Mark II, also known as the B2 [6]. The Paraset continued to be produced by the SOE, but the B2 was commonly used by the agents in the field [4] as it produced a higher output power than the Paraset. The aforementioned SOE agent radio operators, Noor Inayat Khan and Virginia Hall both carried B2 radios (**Pic 3.9**). Its characteristics were as follows:

Receiver

Circuit Features: Mix/Osc, IF Amplifier, IF/BFO Amplifier, Det/AF (AM and CW)

Frequency Coverage: 3.1 - 15.2MHz in 3 ranges:

Range 1	8.7 - 15.2MHz
Range 2	5.2 - 9.04MHz
Range 3	3.1 - 5.4MHz

Intermediate Frequency: 470KHz

AF Output: 50mW into 120 Ohm

Performance: Sensitivity 1 - 3uV for 10mW output (CW)

Valves: 7Q7 (x2), 7R7 (x2)

Schematic: see **Fig 3.4**

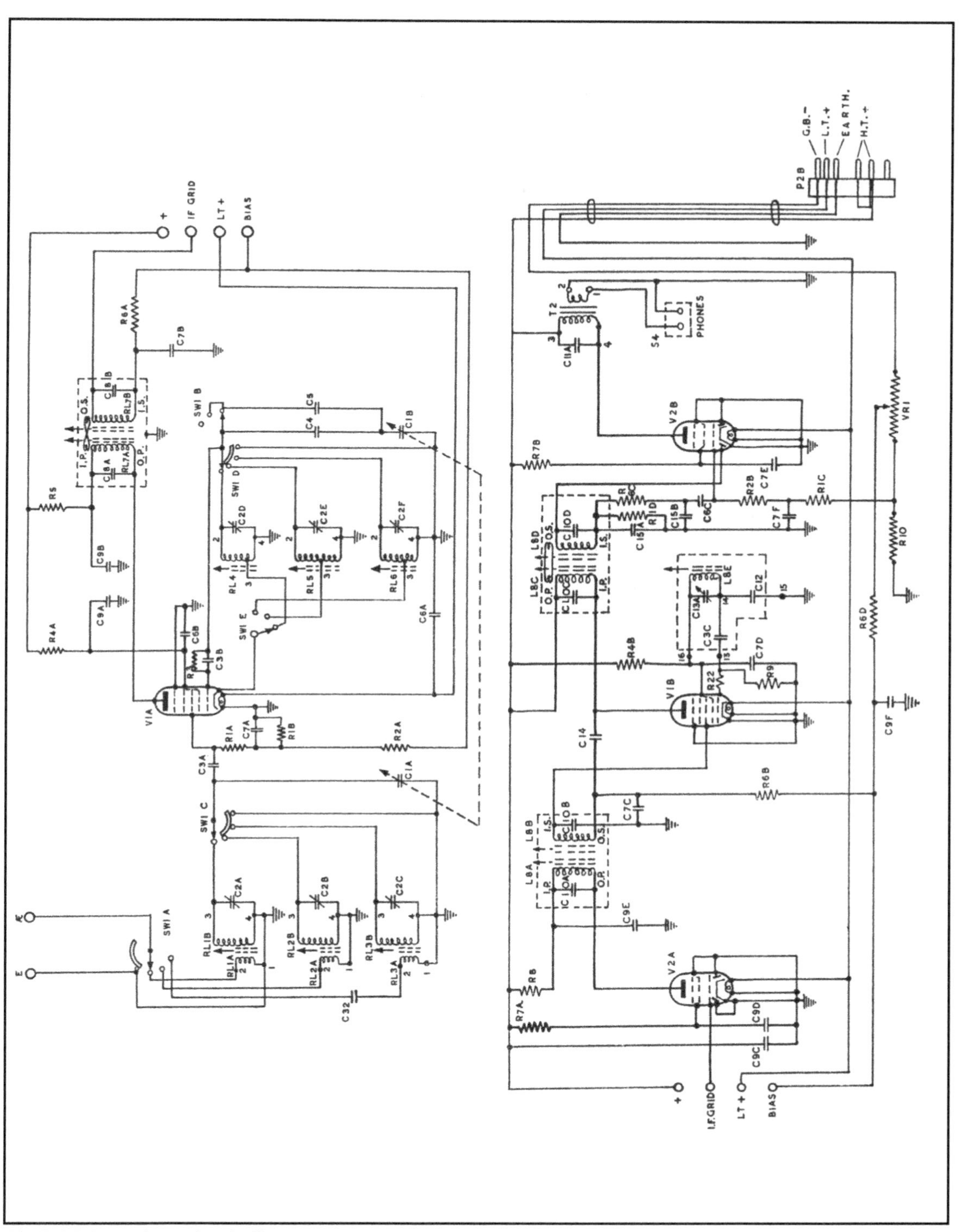

Fig 3.4 B2 receiver circuit diagram

Transmitter

Circuit Features: CO, RF power amplifier (CW only)

Frequency Coverage: 3 - 16MHz in 4 ranges:

Tank coil L1 3.5 - 5MHz

Tank coil L2 4.5 - 7.5 MHz

Tank coil L3 6.5 - 10 MHz

Tank coil L4 9 - 16 MHz

Power Output: Fundamental and second harmonic 20W, third harmonic 16 - 20W

Valves: EL32, 6L6

Schematic: see **Fig 3-5**

Power Supply

AC Mains: 97 - 140V and 190 - 250V, 40 - 60Hz; consumption receive 27W, transmit (key down) approx. 57W

Battery operation: 6V, consumption receive: 4.5A, transmit 9.5A

Individual Units

Receiver 230V at 28mA; 6.3V at 1.2A; - 12.5V bias

Transmitter: 500V at 60mA; 6.3V at 1.1A

General

Size (cm) and Weight (kg):

Item	Height	Length	Width	Weight
Receiver:	12.3	24	11.5	2.3
Transmitter:	12.3	24	16	2.8
Power Pack:	13	27	10	4.6
Spare Box:	13	27	8.5	1.6
Suitcase:	14	32	47	13.2
Antenna:	18m wire and 3m grounding lead			

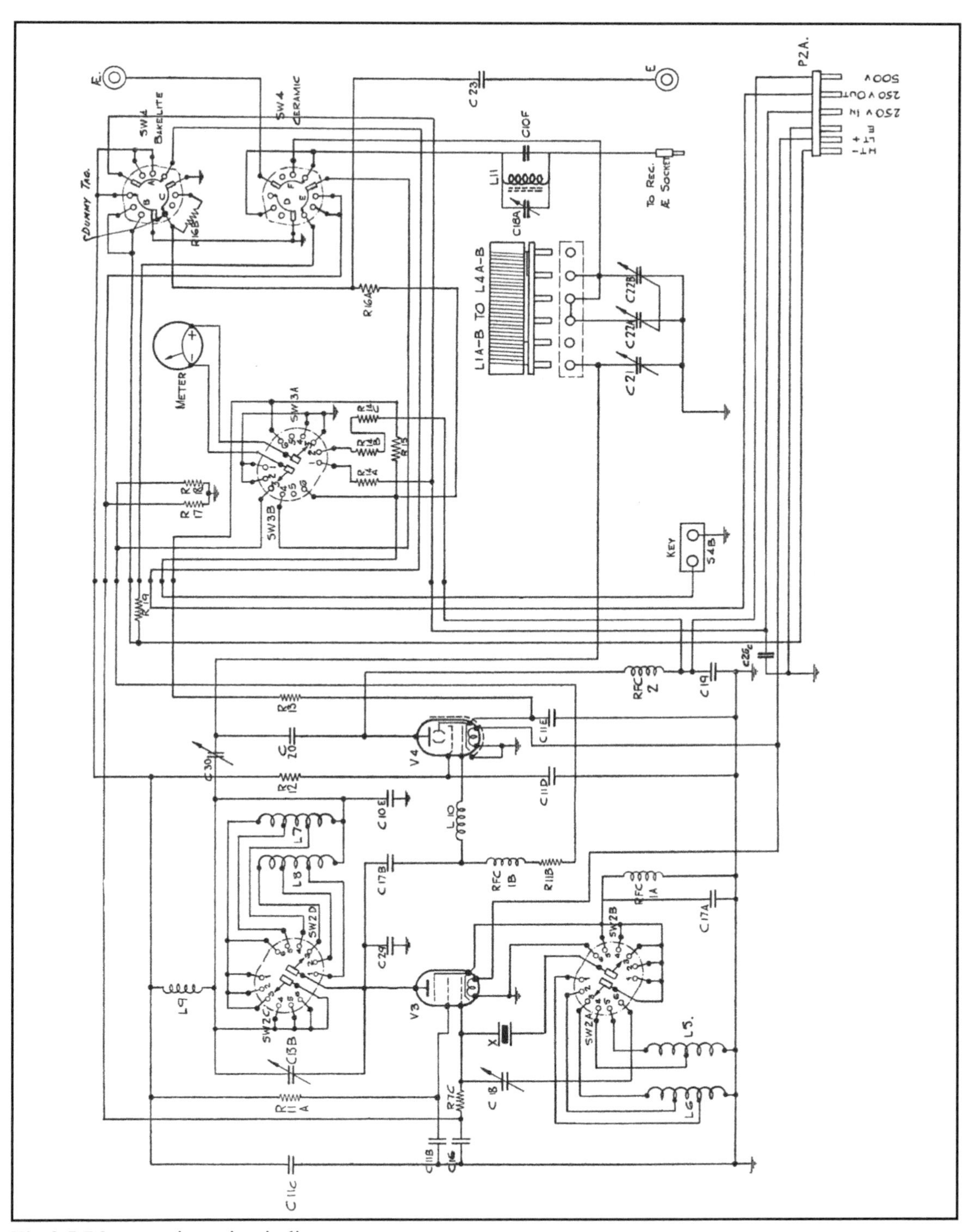

Fig 3.5 B2 transmitter circuit diagram

As can be seen from the photos, the B2, unlike the Paraset, is not a true transceiver in the physical sense, because its transmitter, receiver, and power supply are separate units, which can be carried or stored separately when necessary. Though more powerful than the Paraset, the B2 is much heavier and bulkier.

References

[1] Meulstee and Staritz, *Wireless for the Warrior* Vol. 4. (no pagination)
[2] Strictly speaking, a transceiver consists of an integrated transmitting and receiving sections sharing some of the same circuitry or components, as opposed to an electrically separate transmitter and receiver.
[3] *Wireless for the Warrior* Vol 4. (no pagination)
[4] Dan Petersen, W7OIL, conducted an experiment with his replica Paraset to see how far the spurious signal generated by the regenerative oscillation can travel. He was able to hear it at a distance of three miles.
[5] Translated from the original French and quoted in Jean-Louis Perquin, *The Clandestine Radio Operators,* Histoire & Collections, Paris, France 2011 p. 36.
[6] *Wireless for the Warrior* Vol 4. (no pagination)

4

Building a Replica Paraset

4.1 Assessing the Task Ahead

BUILDING A PARASET replica is not like building an old Heathkit radio, or a modern QRP transceiver from a commercially available kit [1] which simply require a builder to follow a step-by-step instruction from an accompanying manual and, voila!, you have a complete, operable radio right away.

To build a replica Paraset, requires a would-be builder to put together his own parts, obtained from his own, or a friend's junk box; by visiting rallies; searching online stores; or even fabricating his own components. As you see from Pics 4.1 to Pics 4.5, each builder's Paraset replica looks slightly different from the other, even though everyone follows the original schematic and retains the basic philosophy of minimalist construction.

It is quite common that a builder changes a part as a better substitute is found, or is thought of during construction. Even after finishing a working replica, he might go back to replacing some components as he finds a component closer to the original one. It is not that building a Paraset replica is inherently difficult, but it does require patience. I know of no one who has built a Paraset replica in a week or two from start to finish. But many do confess, after it is built, the experience was very rewarding. Fortunately, several builders have shared their experience online, which may help inspire the would-be constructor's approach.

4.2 Getting the Parts – Where to Look

When I mention that I have recently built replicas of a WWII-era spy radio, people, including many Radio Amateurs, often offer an unsolicited comment saying how difficult it must have been to find all the necessary old parts to build one. Some even ask "How can you get valves from the WWII era?" Ironically,

Pics 4.1 to Pics 4.5 examples of replicas built by various Radio Amateurs

Pic 4.2

Pic 4.3

valves such as the 6V6 and the 6SK7, used by the Paraset, are amongst the easiest items to find [2]. Many 'NIB' ('New In Box') and 'NOS' ('New Old Stock') valves, as well as used ones, are consistently available through Internet vendors.

The advent of the Internet has made obtaining many vintage parts much easier than before, but antique radio auctions; Amateur Radio rallies and conventions; and some shops, continue to be good sources for old parts.

• One European vendor in particular focuses on the market for spy-set replications and repair. 'The ParasetGuy' in the Netherlands (email: parasetguy@paraset.nl), offers almost all the parts necessary for construction of a Paraset replica in his catalogue, which includes the front panel chassis, a challenging component to fabricate for most would-be replica builders. His offerings also include the built-in straight key which is another relatively difficult component to duplicate to the original specifications. The proprietor, Henk van Zwam, has an uncanny knack for locating old-stock components of the WWII vintage in warehouses and old factories. He also clones vintage components using modern materials. Through his website, the ParasetGuy offers one of the most useful guides for gathering vintage parts, as well as for actually building a most authentic Paraset replica based on his own experience. https://paraset.nl/aa/?page_id=102

• The Friedrichshafen ham-fest is a very large event for potentially finding that much-needed component.

• In the UK, the British Vintage Wireless Society holds various meets and auctions and may be of help both for tracking-down vintage parts and learning about restoration techniques.

• Farnell is a popular UK source of modern components.

4.3 Safety First!

For the benefit of prospective spy-radio replicators, a brief word of caution is in order. Before proceeding to build and test a valve radio, it is important to be aware that potentially lethal voltages are present. Voltages of 300-400V are in the power supply and on the chassis in a typical valve radio. Many 21st century radio builders are accustomed to working with low voltage solid state devices, but valve radios in general require much higher voltages to operate, and so taking appropriate safety precautions is much more important.

It is particularly important to note that when building a Paraset replica and following the original schematic, that one of the transmitter tank-coil tuning variable capacitors is "hot"; ie the capacitor floats at about 350V - the voltage supplied to the 6V6 anode. You need to make certain that both the rotor and stator sides of that capacitor are electrically isolated from the panel/chassis. Further modifications to avoid exposure to a potentially lethal voltage will be discussed later.

Pic 4.4

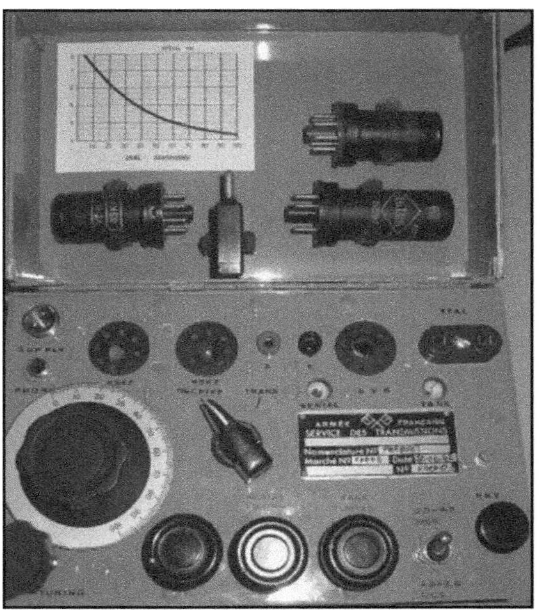

Pic 4.5

4.4 Overview: Mechanical Drawings & Circuit Schematics

The community of the Paraset aficionados owes a major debt of gratitude to Joseph "Joe" Le Suisse, ON5LJ (SK), a Belgian Radio Amateur, who drew a schematic and a precise mechanical drawing of an original Paraset in the 1990s, and made them widely available. This has revolutionised Paraset replication .Other Amateurs, including Jo Scholtes, ON9CFJ; Jean-Claude Buffet, F6EJU; and Johnny Appel, SM7UCZ; have refined Joe's drawings and offer them on the Internet.

The circuit schematic of the Paraset is shown in Pic 4.6 and some fabrication and assembly details in *Pics 4.7 to 4.9.* For more information see [3].

Pic 4.7 to 4.9 Original fabrication and assembly drawings by ON5LJ's

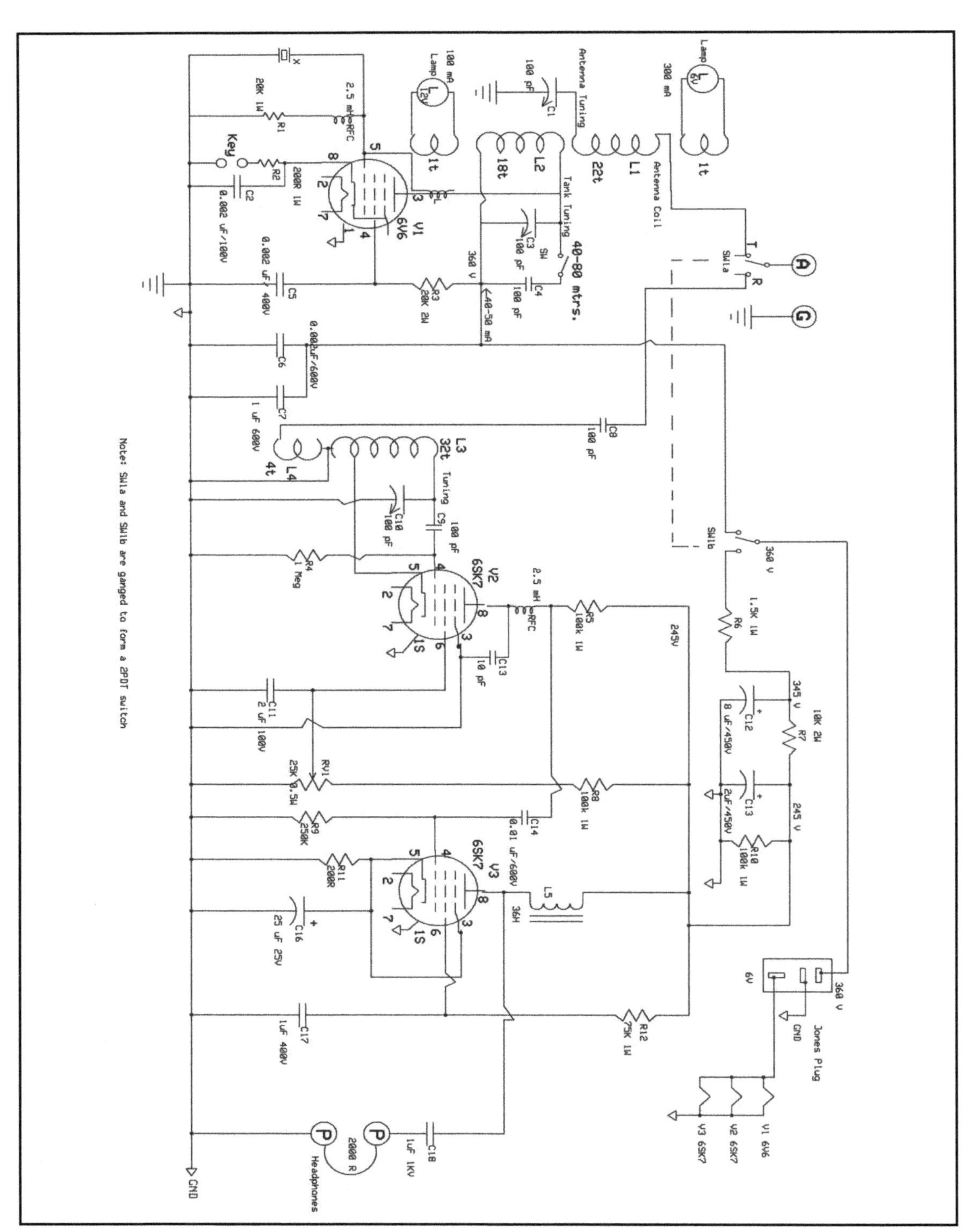

Pic 4.6 Paraset schematic drawn by W3HWJ

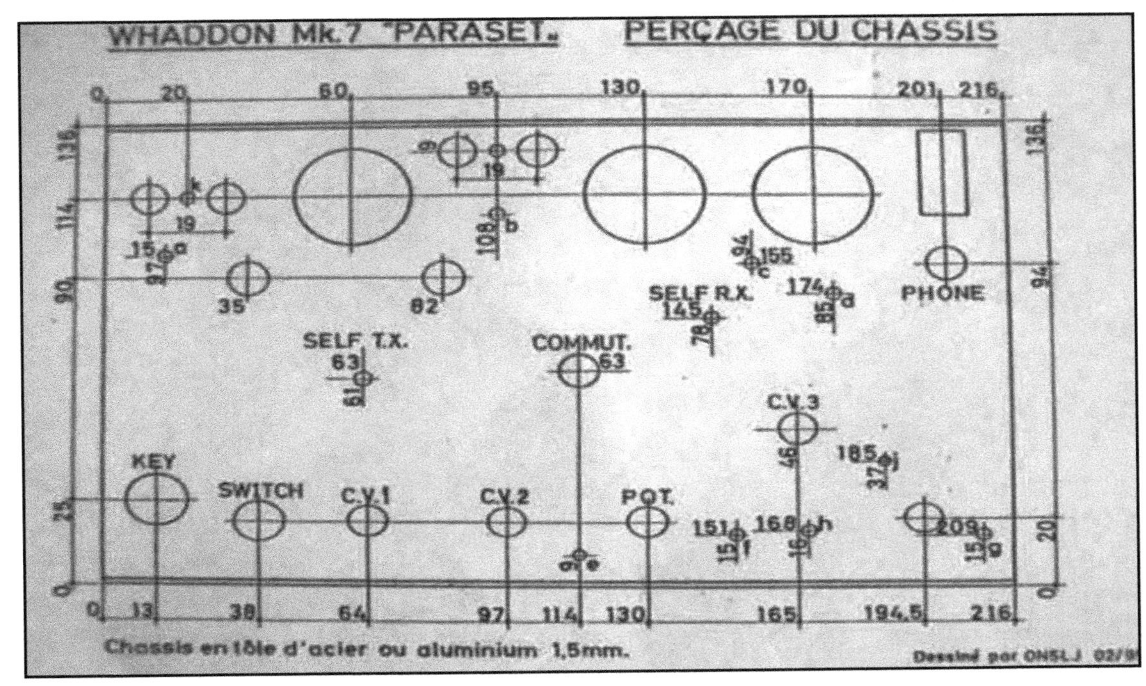
WHADDON Mk.7 "PARASET" PERÇAGE DU CHASSIS
SELF T.X.
COMMUT.
SELF R.X.
PHONE
C.V.3
KEY
SWITCH
C.V.1
C.V.2
POT.
Chassis en tôle d'acier ou aluminium 1,5mm.

Pic 4.7

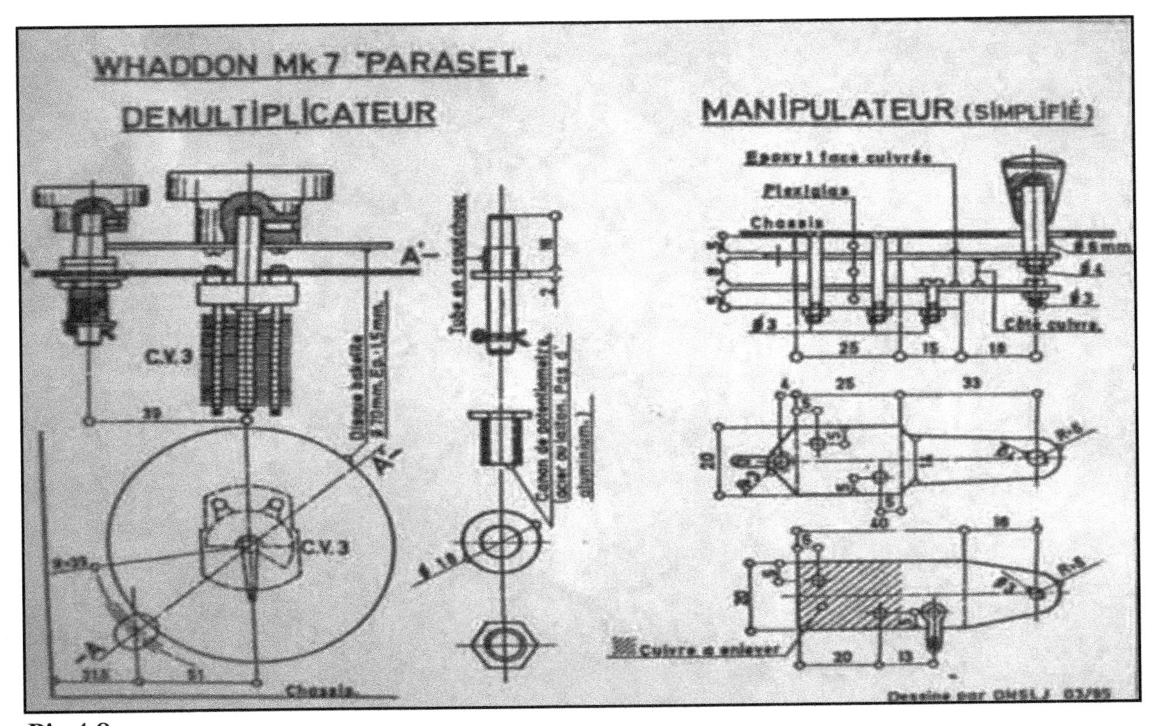
WHADDON Mk 7 "PARASET"
DEMULTIPLICATEUR
MANIPULATEUR (SIMPLIFIÉ)
C.V.3
Chassis

Pic 4.8

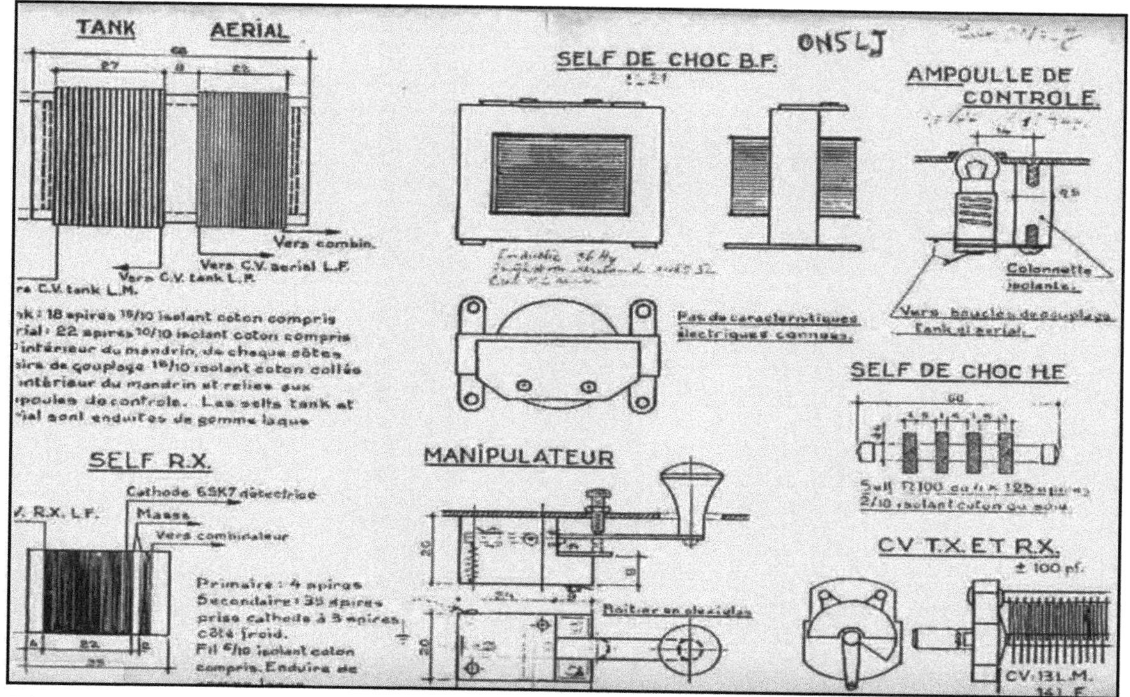

Pic 4.9

4.5 Assembling the Various Parts of the Paraset

The Front panel/chassis

The Paraset's original front panel was made of hard steel, but many replica builders prefer aluminium because it is readily available and easier to shape and drill (Pic 4.10 & 4.11).

Although most builders adhere to the original location of the chassis holes and their dimensions, the outer dimensions of the panel may differ depending on the kind of enclosure the constructor intends to use. However, such variations from the original would not be apparent, except to someone who has carefully studied the original units.

My own first Paraset replica used an aluminum panel that I had on hand and a wooden shoeshine box from an antique shop (Pic 4-13). The box is slightly larger than the original Paraset's wooden box dimensions; hence the panel is made slightly larger than the original specifications in order to fit into this box.

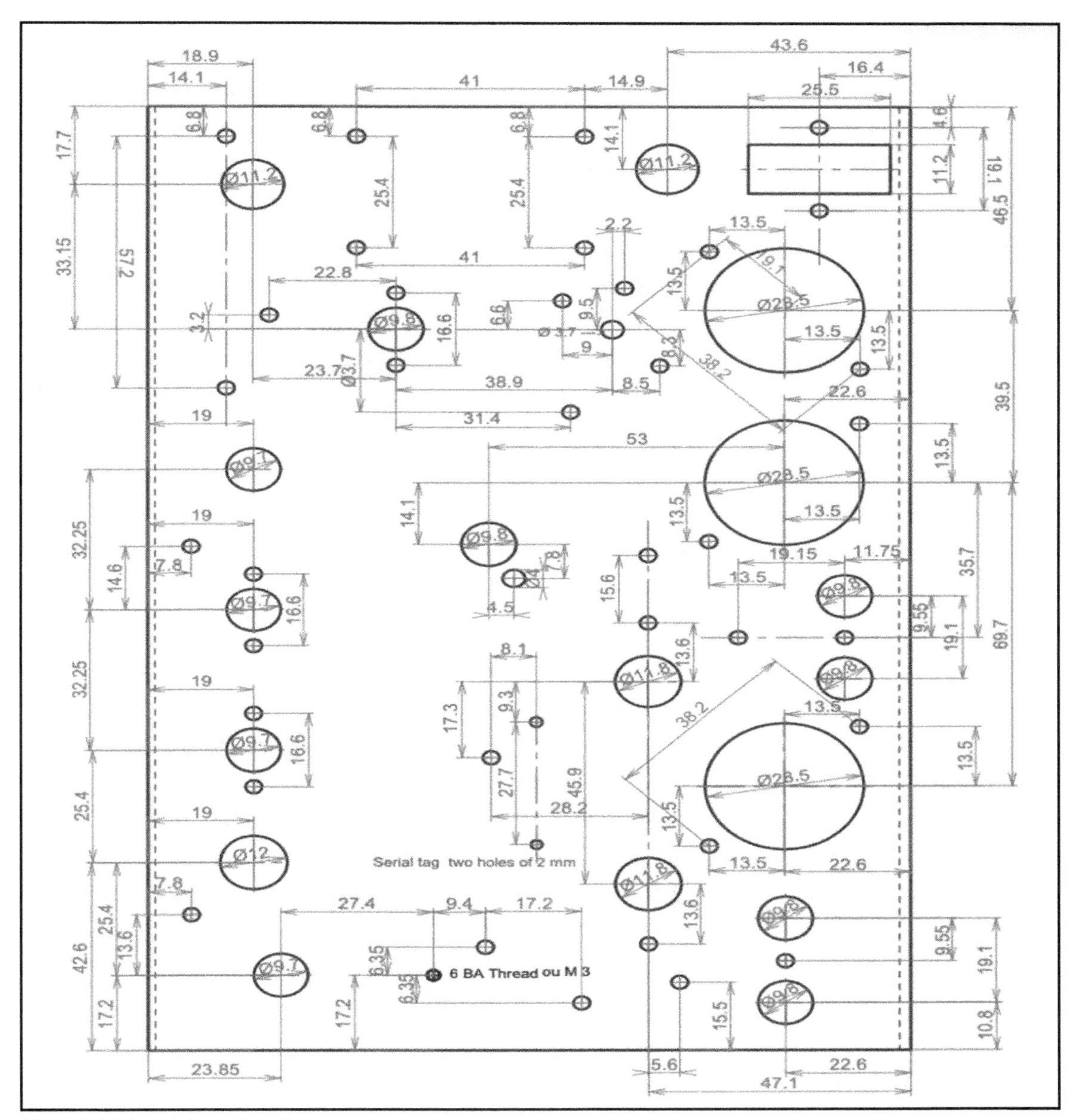

Pic 4.10 Machining drawing

2. The Enclosure

Early Parasets were built in a wooden enclosure, and carried in a leather suitcase, such as the one on display at the Imperial War Museum in London (Pic 4.14). A later version was built in an all-metal 'cash box' case, intended to reduce RF signal leakage and lower the risk of detection (Pic 4.15). When

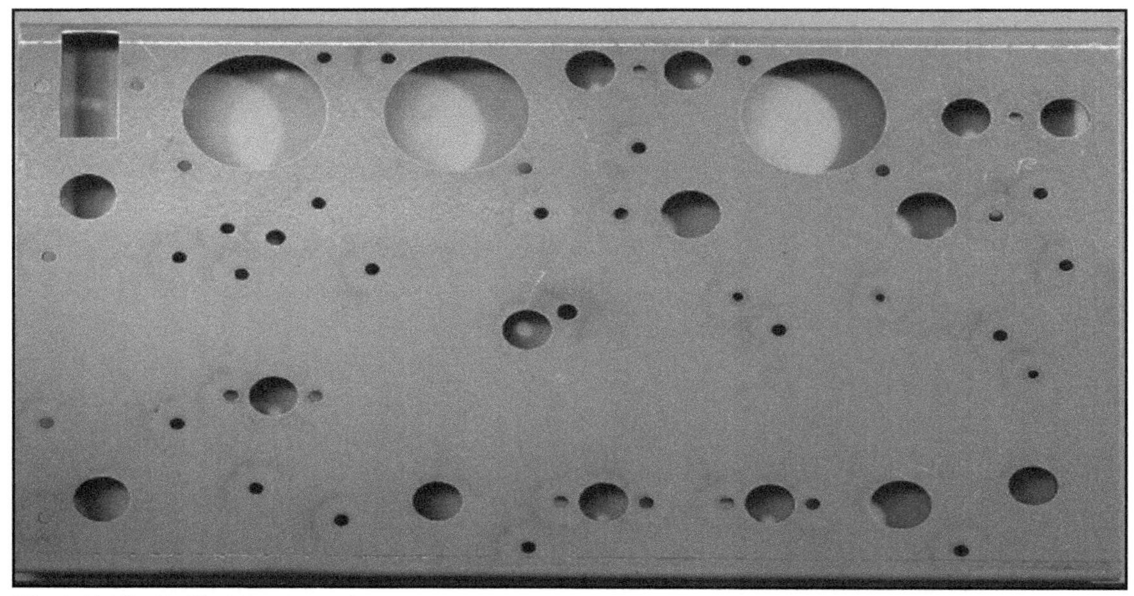

Pic 4.11 The Drilled front panel

Pic 4.12 A replica built by WA8YWO

Pic 4.13 The Author's replica Paraset

closed, and while not in operation, the radio looks like an ordinary utility box, which might be commonly found in a kitchen or on a shelf.

Dimensions for the 'cash box' enclosure are shown in Pic 4.16.

Peter Jensen, VK7AKJ, describes, step-by-step, how he constructed his 'cash box' utilizing galvanized steel in his book, *Wireless at War* [4] (Pic 4.17).

For further information about fabricating a front panel/chassis, as well as a metal "cash box" enclosure construction, see: http://www.sm7ucz.se/Paraset_F6EJU/Paraset_F6EJU.htm

Pic 4.14 Original Paraset in leather/wooden case

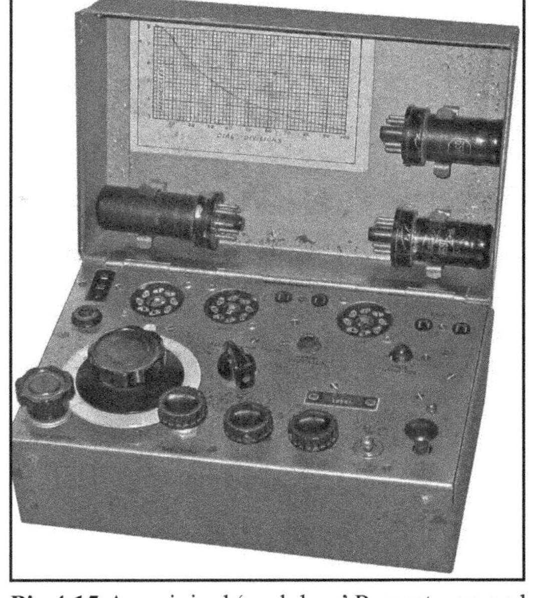

Pic 4.15 An original 'cash box' Paraset - owned by Mark Meltzer, AF6IM

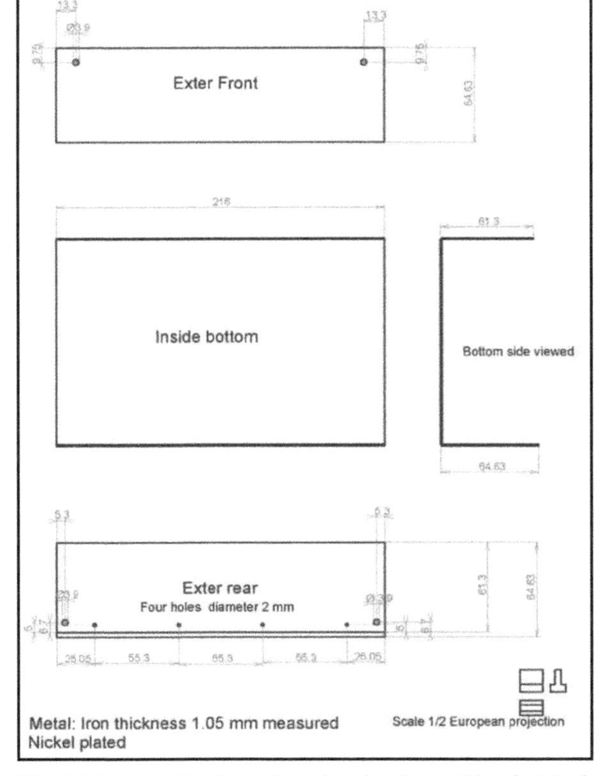

Pic 4.16 A meticulous drawing by Jean-Claude Moffet, F6EJU, based on his measurements of the original Paraset metal box enclosure.

Pic 4.17 Peter Jensen's 'cash box' Paraset

Pic 4.18 National friction-drive

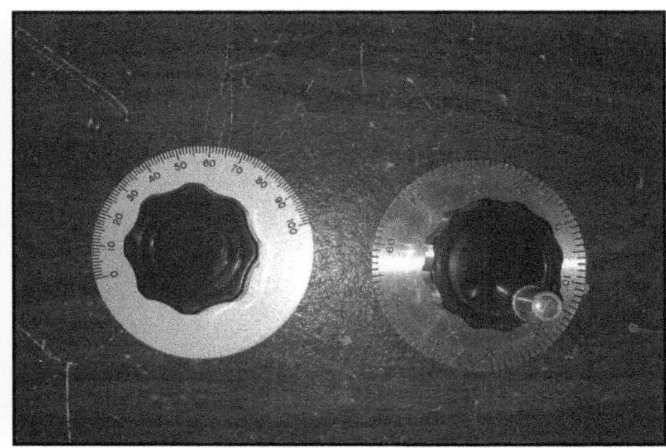

Pic 4.19 Dials which will serve the purpose

3. The Tuning Dial

The large tuning dial (with knob and friction-drive mechanism) is one of the few relatively difficult-to-find vintage parts with which to build an authentic replica. Manufacturers such as National formerly made a suitable friction-drive mechanism (Pic4-18), but the author knows of no manufacturers who produce this kind of drive today. However, several dials close in appearance and size to the original have been acquired by the author from rallies and swap meets over the past ten years (Pic 4-19), which suggests that, unless you are lucky enough to find a vintage one, you will need to fashion your own.

→To fabricate an authentic friction drive (Pic 4.20 & Pic 4.21) see:

http://www.sm7ucz.se/Paraset/lilldrevet_1.jpg;

http://www.sm7ucz.se/Paraset/Paraset_friction_drive_e.htm

Pic 4.20 An authentic friction drive by Johnny, SM7UCZ

Pic 4.21 An authentic friction drive by Johnny, SM7UCZ

Pic 4.22 a home-made dial and friction-drive by Henk of ParasetGuy

Pic 4.23 a home-made dial and friction-drive by Henk of ParasetGuy

A further example of a home-brew dial and drive is shown in Pic 4.22 & Pic 4.23:

Although less authentic, one can substitute a 2 - 3 inch diameter Vernier dial, as Michael Tyler, WA8YWO did (Pic 4.26) without really compromising the minimalist 'spirit' of the Paraset.

Alternatively, instead of the drive, one could simply install a 10 pf variable capacitor for fine tuning, connecting it in parallel to the tuning capacitor as Dan Peterson, W7OIL, has done.

Pic 4.24 Substituting a Vernier dial

4. The telegraph (straight) key

Jean Claude, F6EJU, offers accurate drawings and instructions on how to build the key (Pic 4.25 – Pic 4.27).

Dan, W7OIL, built his key using easily available material from a local hardware shop (Pic 4.28).

You can also use a micro-switch; or a leaf-switch blade with insulating material from an old large toggle switch, as I did for my third replica (Pic 4.29 - Pic 4.31).

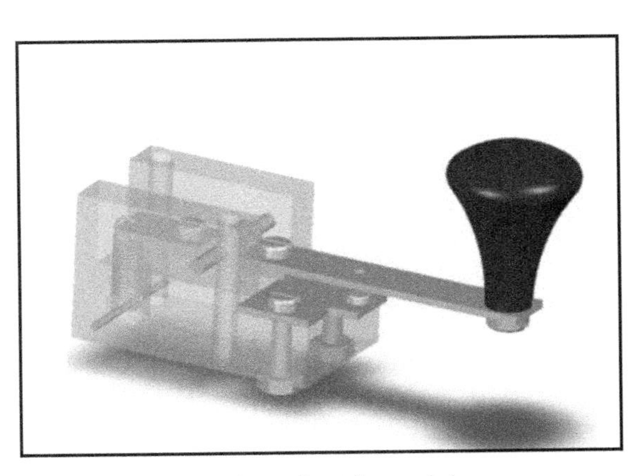

Pic 4.25 Construction of a telegraph key

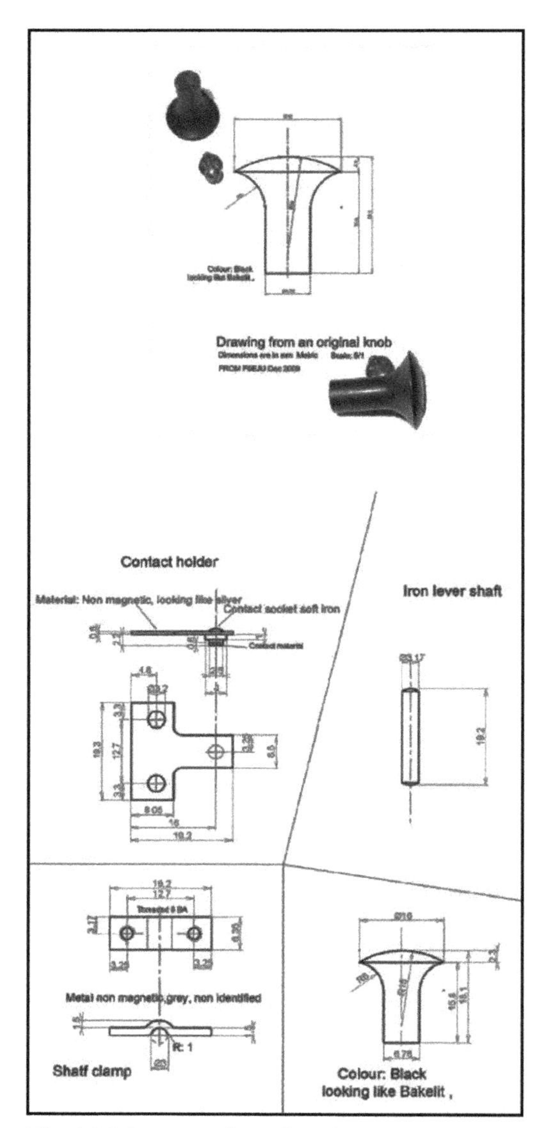

Pic 4.26 Construction of a telegraph key

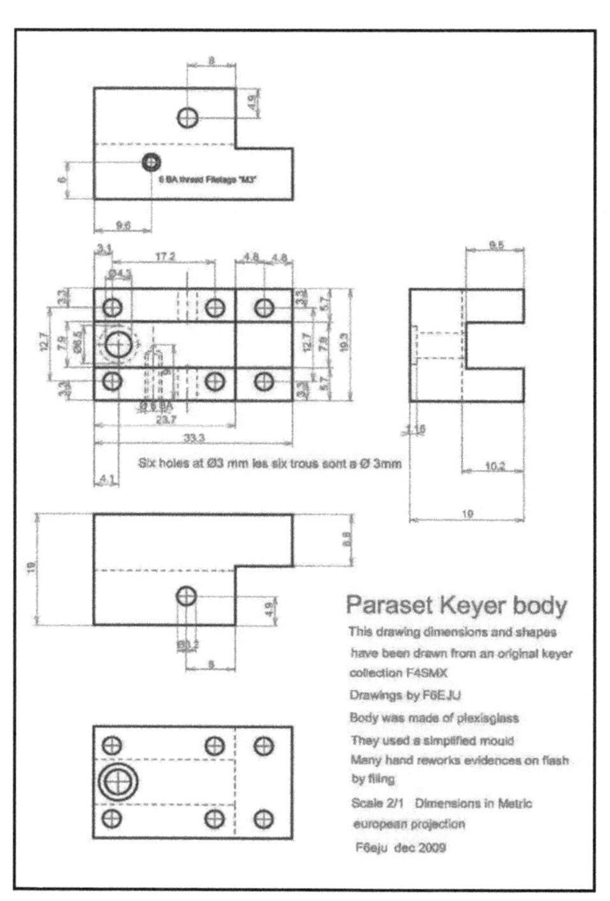

Pic 4.27 Construction of a telegraph key

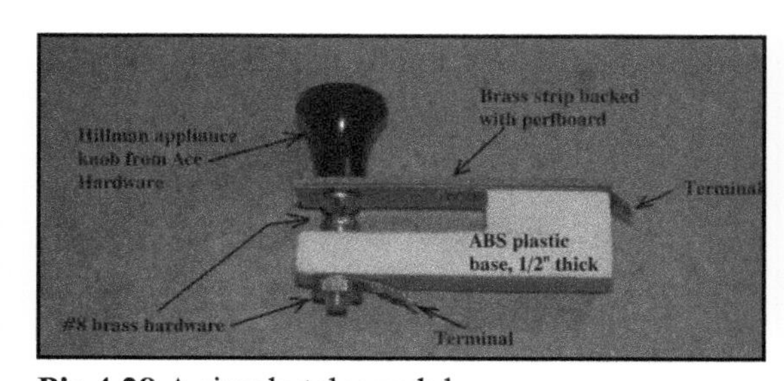

Pic 4.28 A simple telegraph key

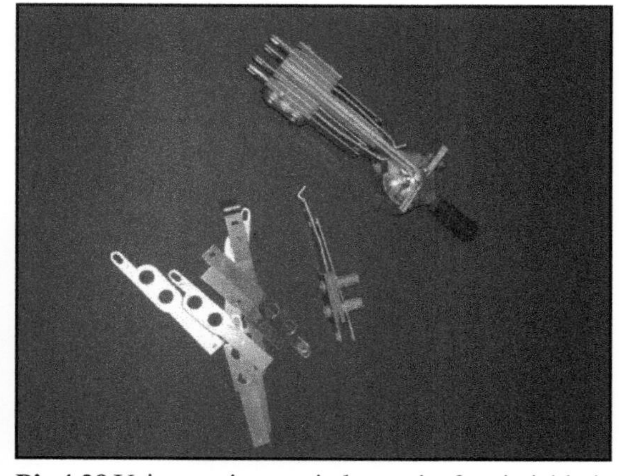

Pic 4.29 Using a micro-switch; or a leaf-switch blade with insulating material, as a key

Pic 4.30 Using a micro-switch; or a leaf-switch blade with insulating material, as a key

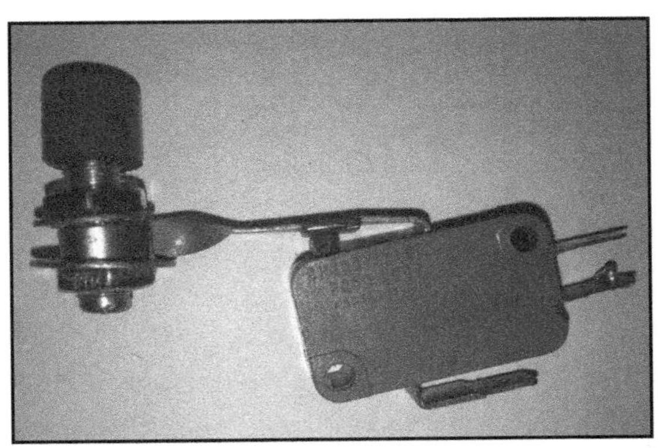

Pic 4.31 Using a micro-switch; or a leaf-switch blade with insulating material, as a key

As an aside, there is an interesting true story of how to make the simplest key for a clandestine radio by using a nail, a kitchen knife and a book. Lily Serguief, was a double-agent pretending to work for the Abwehr in Paris, but in fact was sending disinformation from London to German intelligence. She was issued with a clandestine radio, and was instructed by her Nazi handler, Major Emile Klieman's, a radio technician, on the eve of her departure from Lisbon to London:

"Here are the two holes to which you connect the Morse tapper. It's only outward sign that could arouse an expert's suspicions. Of course, we won't give you an actual [key] to take with you: it would be too risky. But I'll show you how to make one. It's very easy: you take a small wooden board and pierce it with a nail so that the point sticks out about half an inch. Before you drive the nail home, you fix one wire of an electric light flex round it. You place your board down with the point of the nail upwards. On the board you put a book with the blade of a kitchen knife slipped between the pages. You attach the second wire of the flex to the blade of the knife and you plug the other end of the flex into a wireless set, where I showed you. You must make sure the knife blade is held firmly in the book, and that the handle-end sticks out, just above the point of the nail - then all you have to do is to press on the handle to make a contact." [5]

5. Coils, and coil formers

Winding information for the Paraset coils is shown in Pic 4.32

As to obtaining the former on which to wind the coil, Amateur radio meets are a good source, but otherwise, a PVC pipe from a local hardware shop, a pill container, or a plastic medicine bottle will do nicely if you can't find them (Pic 4.33 & Pic 4.34). Copper wire is available from most hardware shops,

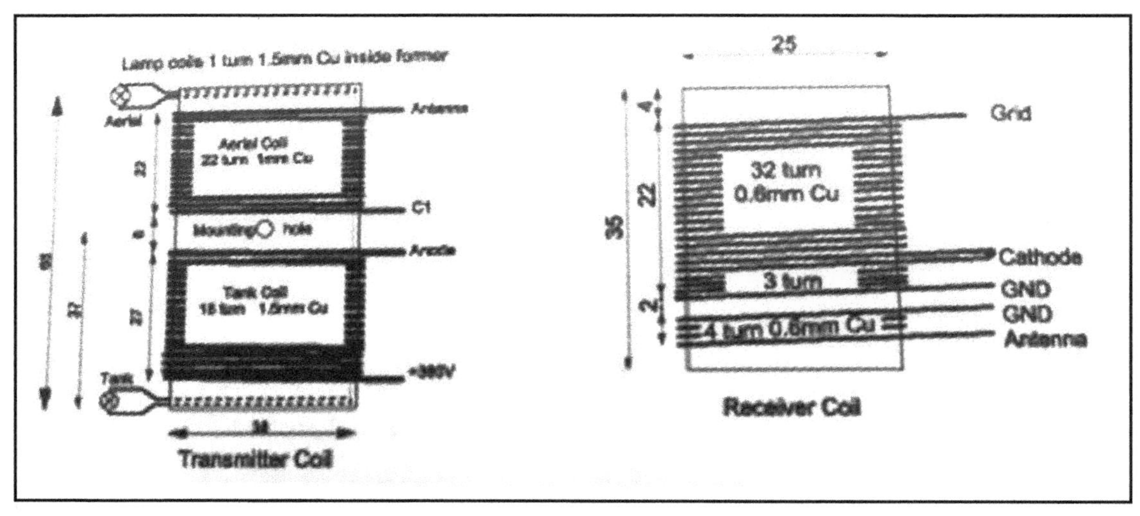

Pic 4.32 Paraset coil winding information

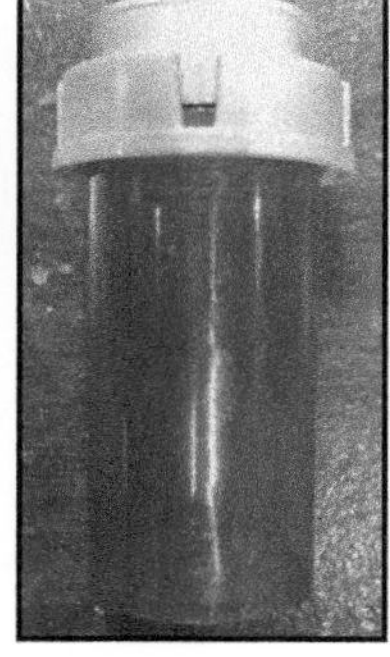

Pic 4.33 & Pic 4.34 Types of coil former that can be used

Pic 4.35 Coil Winder by Henk of ParasetGuy

or one may be able to cannibalize some from an old broadcast or short-wave radio.

Although one can wind coils by hand, to make life easier, some replica builders have made a nice winding machine using Meccano (Pic 4.35).

6. Front Panel Knobs

Finding the right four knobs might make your visit to an Amateur meet worthwhile (Pic 4.36). You may not be able to find the exact vintage knobs but there are many available that are close enough in size, shape, and colour.

7. Crystals

The Paraset used two kinds of crystals: a round-headed Bliley type and a square-case type with the same pin size and pin separation. Their leg spacing is wider than the more common FT-243 crystals used by Radio Amateurs in pre-1970s era. Some replica builders have used an FT-171 type crystal which has fatter pins than Bliley's. The four types of crystal are shown in Pic 4.37.

The Bliley and square-type crystals used in the original Paraset are difficult to find at a reasonable price today, especially those that work readily on Amateur band frequencies. Many Radio Amateurs who have replicated the Paraset have used FT-243 or FT-171 types, even though they are not an authentic match. If you have Bailey type, or square-type crystals but not on Amateur band frequencies, you can easily remove the original non-Amateur frequency crystal element from inside, and replace it with a cheaply available Chinese-made HC49 type crystal as you see in this photo (Pic 4.38 & Pic 4.39).

Pic 4.36 Knobs for the Paraset

Pic 4.37 Types of crystal

Pic 4.38 Replacing the crystal element

Pic 4.39 Replacing the crystal element

Pic 4.40 Crystal adapter

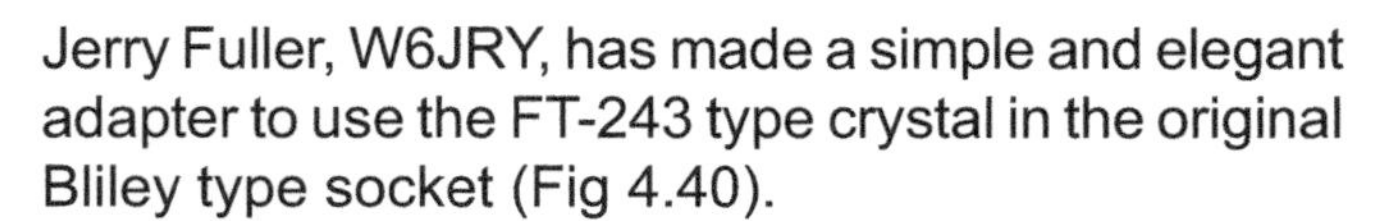

Jerry Fuller, W6JRY, has made a simple and elegant adapter to use the FT-243 type crystal in the original Bliley type socket (Fig 4.40).

8. The 3-prong Cinch-Jones power connector.

Pic 4.41 shows a picture of the power connector used in the original Paraset. Functionally, of course, any connector with three or more separate contacts will do. Many Radio Amateurs use a four-prong Jones type connector which is more commonly available. A cursory look at the eBay market at the time of writing yielded several appropriate three-prong Cinch-Jones connectors.

The original Paraset employs a female connector on the panel, but this exposes a high voltage from the power supply as one inserts the male connector to the female side. To avoid this potential risk, one can simply reverse the installation: the male connector on the chassis and the female connector from the power supply as you see in the photos.

Pic 4.41 Cinch-Jones connectors

9. The Headphone jack

The original looks like the one on the right of Pic 4.42. ParasetGuy.nl still makes these available, but any 1/4" phone jack connector will do the job without altering the look.

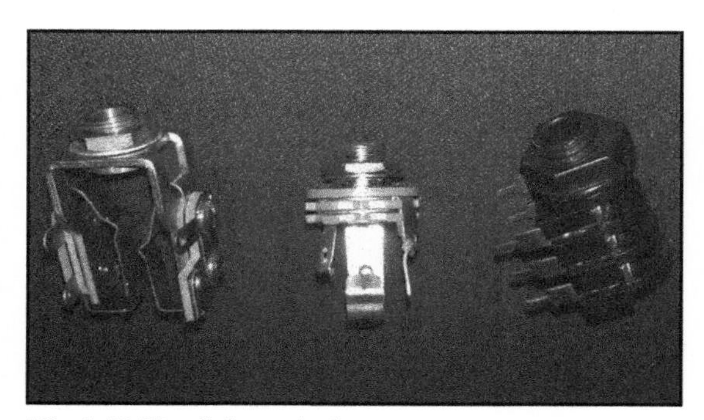

Pic 4.42 Headphone jack

10. Headphones

These are high impedance (2000 - 5000 ohm) and are still abundantly available on eBay and elsewhere (*Fig 4.45)*.

11. Variable capacitors

You will need three 100 pf variable capacitors capable of being isolated from the

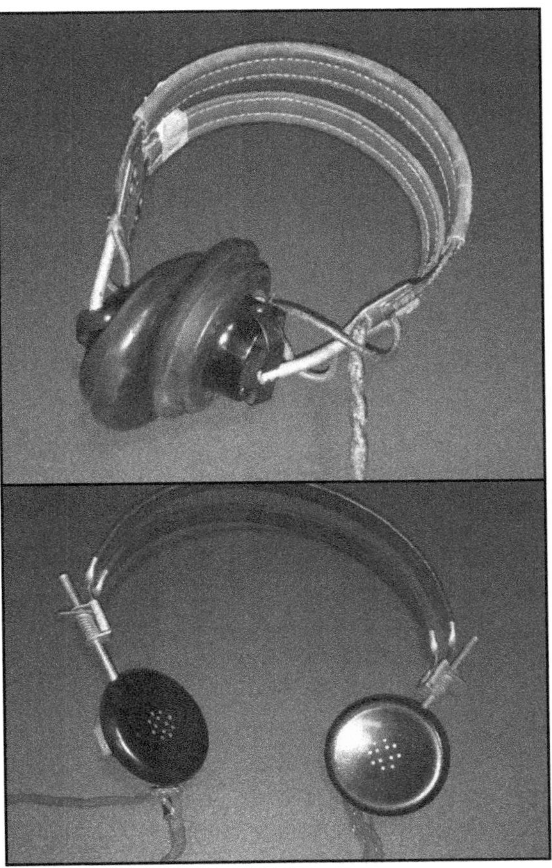

Pic 4.43 Headphones

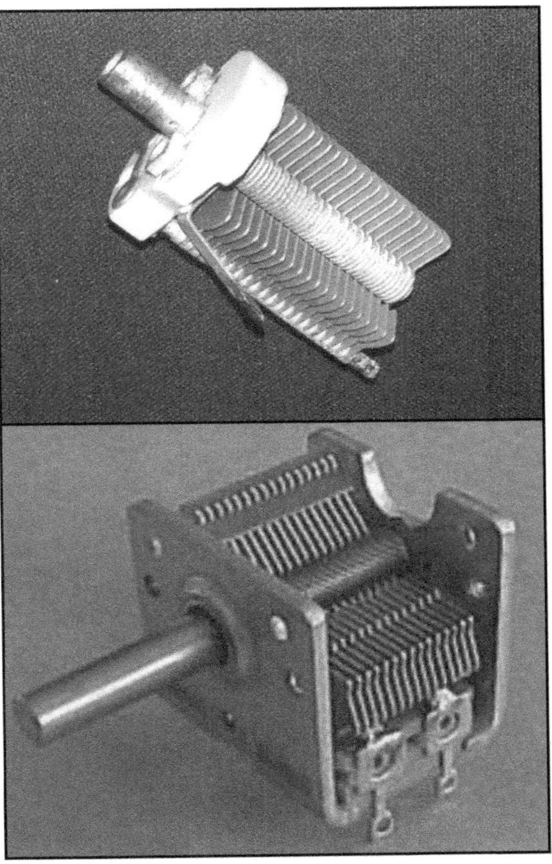

Pic 4.44 & 4.45 Variable capacitors

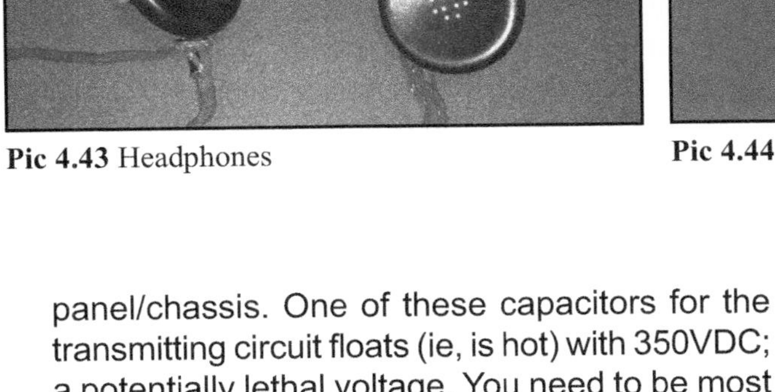

panel/chassis. One of these capacitors for the transmitting circuit floats (ie, is hot) with 350VDC; a potentially lethal voltage. You need to be most careful how you mount it to the panel/chassis in order to insure that the voltage is not conducted on the chassis, exposing the user to the voltage. The original type of capacitors have become more and more difficult to find, so you may need to substitute more commonly available variable capacitors (Fig 4.47, Fig 4.48).

A modification to prevent exposing the high voltage at the variable air-spaced capacitor used in the transmitter circuit has been devised by

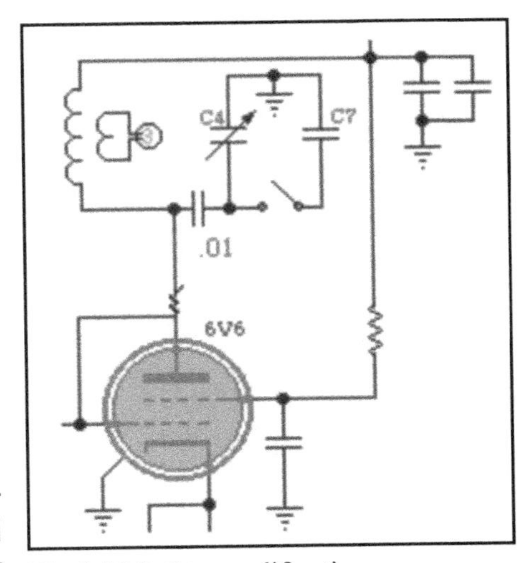

Pic 4.46 Safety modification

Pic 4.47 Original Choke

Pic 4.48 Replacement 60H Choke

Steve McDonald (Pic 4.46). The value of this DC block (safety) capacitor is 0.01uF; working voltage 500v or higher.

12. 30H audio choke

An exact substitute audio choke is very difficult to find these days (Pic 4.47). Some Radio Amateurs have added a cosmetic touch to make a same value choke look like an original. The Canadian company, Hammond, manufactures a 60H choke (also available from ParasetGuy.nl) which works just fine (Pic 4.48). Another approach is to adapt an old speaker transformer by using only the primary side, as I have done with my first replica. The value is not critical. Any value above 15H, or a DC resistance higher than 500 ohm will work.

13. The Rotary Switch

The 2-pole, 2-position rotary switch is an easy item to find on the Internet and elsewhere (Pic 4.49).

14. On-off switch

The short-throw single-pole, single-throw switch manufactured by Gardner Bender Inc (Part #GSW-125) and widely available, is as close to the original switch as can be found today (Pic 4.50).

Pic 4.49 Rotary Switch

Pic 4.50 Toggle switch

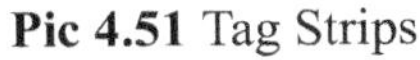

Pic 4.51 Tag Strips

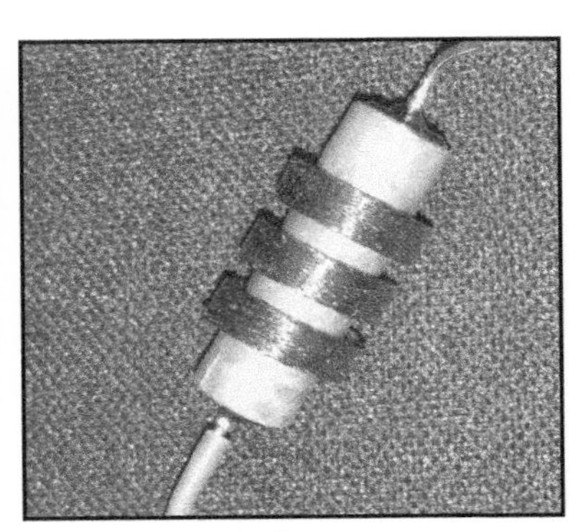

Pic 4.52 RF choke

Pic 4.53 Octal valve sockets

15. Tag Strips

4 off 3-position, and 1 off 7-position; these are very commonly found in vintage valve radios (Pic 4.49).

16. RF chokes

Two 2.5 mH chokes are required, rated at 100 mA or higher (Pic 4.52).

17. Octal valve sockets

These can be salvaged from old valve radios. Three are required (Pic 4.53).

Pic 4.54 Antenna/ground connectors

18. Antenna/Ground connectors

Original antenna and ground connecting posts (Pic 4.54) are nearly impossible to find today, but there are many other types of suitable vintage connectors on eBay and elsewhere.

The one shown in Pic 4.55 is available from ParasetGuy, although the easiest substitute may be a 3-way (or 5-way) audio binding post using the type shown in Pic 4.56.

Pic 4.55 Antenna/ground connector obtained from ParasetGuy

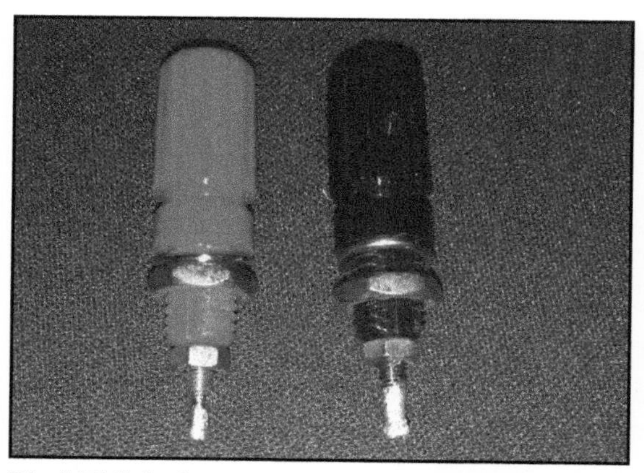

Pic 4.56 Substitute connector using commonly available audio binding posts

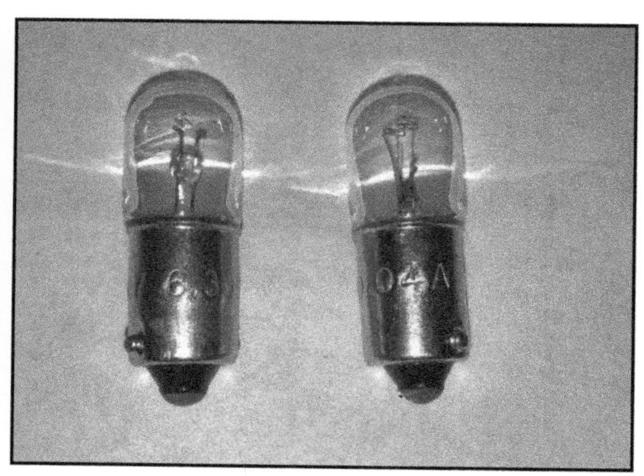

Pic 4.57 Dial bulbs

19. Dial Lights

Two 6.3V dial bulbs are required. You can find these in old broadcast radios and amateur radio equipment. Make sure you also have sockets for the bulbs, although you can directly solder wire to the bulbs, too (Pic 4.57).

20. Capacitors and resistors

You can look for vintage parts, but old capacitors especially, and electrolytic ones in particular, are notoriously dysfunctional, so make certain they are useable before you solder them in. Modern and more reliable equivalent

Item	Farads	Volts	Type (fixed unless otherwise stated)
C13	10 pF	300	
C9	100 pF	100	
C8	100 pF	100	
C1	100 pF	variable	
C3	100 pF	variable	
C10	100 pF	variable	
C4	100 pF	600	
C2	0.002 μF	100	
C5	0.002 μF	400	
C6	0.002 uF	600	
C7	1μF	600	
C17	1μF	400	
C14	0.1 μF	400	
C18	0.1 μF	1000	
C11	2μF	100	Electrolytic
C15	2μF	450	
C12	8μF	450	Electrolytic
C16	25μF	25	Electrolytic

Table 4.1 Capacitors

parts are easily available online or in electronic supply shops. The reference numbers refer to the schematic in Pic 4.6.

You can use a higher voltage-rated capacitor, for example, instead of 1 µF 400V, you can use 1 µF 600V.

Item	Ohms	Watts	Type
R	200	1	carbon
R11	250	1	carbon
R5	1.5 k	1	carbon
R7	10 k	2	carbon
R1	20 k	1	carbon
R3	20 k	2	carbon
R12	75 k	1	carbon
R6	100 k	1	carbon
R8	100 k	1	carbon
R10	100 k	1	carbon
R9	250 k	1	carbon
R4	1 M	0.25	carbon

Table 4.2 Resistors

In the original Paraset, all resistors are of carbon type, but any type of resistor will work as long as their power rating in watts is as designated or higher.

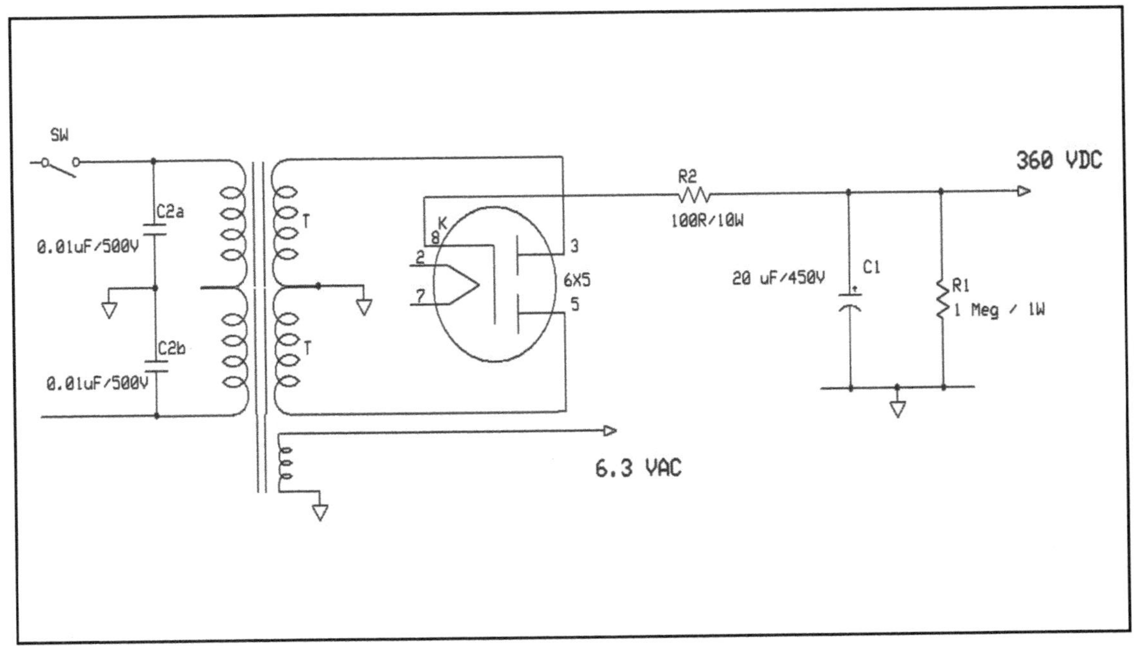

Pic 4.58 Schematic of Paraset AC Power Supply drawn by W3HWJ

Pic 4.59 Original AC Power Supply – external view

Pic 4.60 The finished AC power supply – external view

21. Power supply

The original Paraset used two kinds of power supply: an AC mains input unit and a battery input unit using a vibrator. The schematic for the AC mains input power supply, and how the original looked, are shown in Pics 4.58 and 4.59 respectively.

I built my replica with a diode-bridge rectifier instead of a 6X5 valve (Pic 4.60 and 4.61 respectively).

Pic 4.61 The finished AC power supply – internal view

4.6 Testing the Paraset

6SK7 RF Stage

Valve Pin No.	Min Regeneration	Max Regeneration	
Pin 1 Metal	0V	0V	
Pin 2 Heater	0V	0V	
Pin 3 Grid 3	0V	0V	
Pin 4	Grid 1	-0.52V	-0.43V
Pin 5	Cathode	0V	0V
Pin 6	Grid 2	0V	27V
Pin 7	Heater	6.5V	6.5V
Pin 8	Anode	213V	119V

6SK7 Audio Stage

Pin 1	Metal	0V
Pin 2	Heater	0V
Pin 3	Grid 3	2.3V
Pin 4	Grid 1	-
Pin 5	Cathode	2.3V
Pin 6	Grid 2	74V
Pin 7	Heater	6.5V
Pin 8	Anode	203V

6V6 Transmitter Stage

Pin 1	Metal	0V
Pin 2	Heater	0V
Pin 3	Anode	323V
Pin 4	Grid 2	243V
Pin 5	Grid 1	-
Pin 6	N.C.	
Pin 7	Heater	6.5V
Pin 8	Cathode	9.36V

Table 4.3 Measured voltages at valve pins (Henk van Zahm)

4.7 Modifications

Although the purist may object to any change to the original design, there are ways to improve the safety and usability of the Paraset without altering the appearance of the original:

* **Important:** To avoid exposure to high voltages, reverse the Cinch-Jones male/female connectors that connect to the front panel (see Section 4.4 (8). For a similar reason, fit a DC blocking capacitor between the anode tuning variable capacitor and the transmitter HT rail (see Section 4.4 (11)).

* To provide a 'spotting' capability, replace the two-pole, two-position rotary

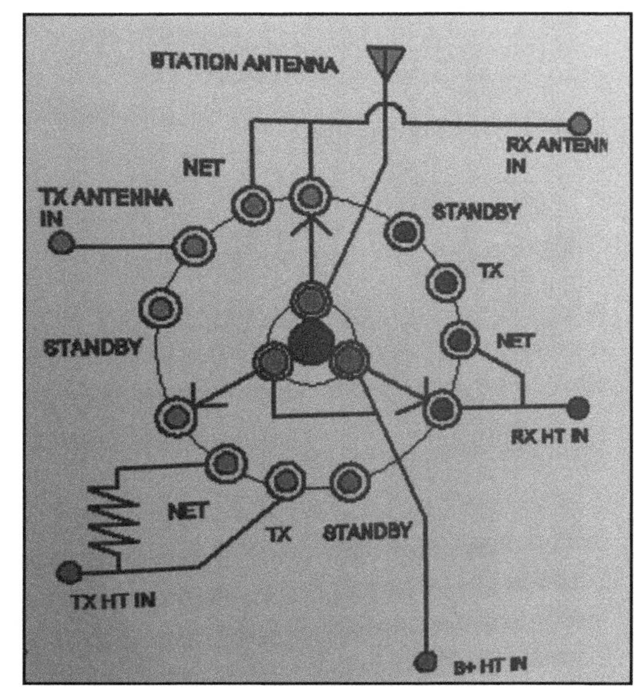

Pic 4.62 Adding a spotting capability

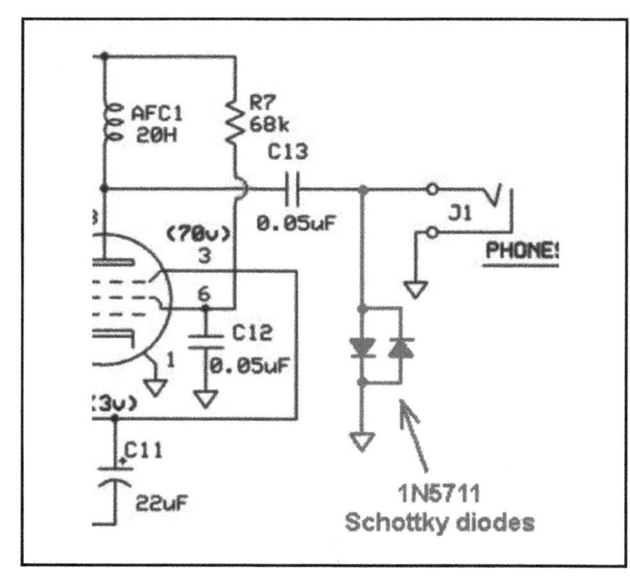

Pic 4.63 Click suppression modification

switch (as employed in the original to switch between transmit and receive), with a four-pole, four-position rotary switch to enable four states of the transceiver: STANDBY-TRANSMIT-SPOT-RECEIVE (Pic 4.62). [6]

* To prevent the loud click that can occur when switching from transmit to receive (a known and irritating characteristic of the Paraset!), add two back-to-back 1N5711 Schottky diodes to the audio output circuit (*Pic 4.63*). [7]

4.8 Some who have Successfully Replicated the Paraset

A number of Radio Amateurs have built their own superbly crafted working replicas and have shared their experiences widely. These generous souls have distributed helpful information and hints on the Internet, in magazine articles, as well as through personal appearances at local amateur radio club events. The following are but a few of them. While this is certainly not an exhaustive list, it should be a starting point for the reader to explore further. I am sure there are others, and I mean no slight to any of the other spy-radio constructors who so generously have contributed to the preservation of these instruments. In appropriate places in this chapter I have drawn on the experience of many of these:

Steve McDonald, VE7SL, on his website, not only chronicles how he built his Paraset replica, including how he repurposed old components (his front-panel/chassis is made of a salvaged STOP sign metal sheet!) but also offers a few ways to improve the Paraset's performance and usability by modifications to the original circuit. He also shares the story of his Paraset contact between British Columbia and South Africa. https://qsl.net/ve7sl/paraset.html

Jerry Fuller, W6JRY, is a superlative craftsman whose replica has the most "professional" appearance of all the replicas I have seen. He is a perfectionist in all things he undertakes, and his Paraset reproduction - he has built four of them! - is so authentic that at a glance it cannot be easily distinguished from the original (Pics 4.64 & 4.65). Incidentally, after I visited his shack for the first time in early spring 2018, he and I had a Paraset-to-Paraset contact on the 40 meter band between our homes, which are 200 miles apart in California. Here is his replica:

Pic 4.64 Jerry Fuller's Paraset replica – top view

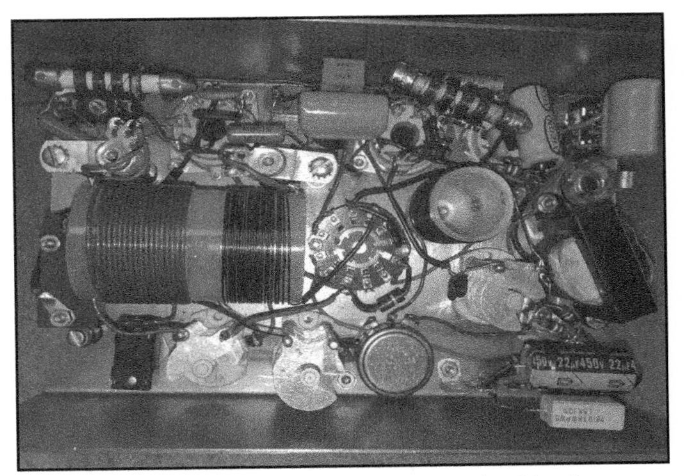

Pic 4.65 Jerry Fuller's Paraset replica – underside view

Graham Stannett, G4VUX has an excellent YouTube presentation available showing how he built his Paraset replica. His video is easy to follow, and would be a most helpful site for any would-be replica builder. https://www.youtube.com/watch?v=JwMc53nE9kQ

Peter R. Jensen, VK7AKJ, in Australia, also chronicles how he built his Paraset replicas, and shares many useful experiences, some which are mentioned in this chapter.

See also these helpful websites:

https://www.qsl.net/ik0moz/paraset_eng.htm.
http://sm7ucz.se/Paraset_F6EJU/Paraset_F6EJU.htm
http://www.sm7eql.se/paraset/index.htm
https://paraset.nl/aa/?page_id=102
http://www.theparasetclub.co.uk/
https://qsl.net/ve7sl/paraset.html

At the time of writing there is one active group of Paraset replica builders who share opinions, advice and information on the Internet: ParasetBuilders@yahoogroups.com.

In the next chapter, we look at using solid-state components, instead of valves. Happy Soldering!

References

[1] Such as the QCX transceiver offered by QRP Labs* or the Para40/80set available from QRPGuys** in California.
*https://www.qrp-labs.com/qcx.html
** https://qrpguys.com/
[2] At the time of writing, eBay lists a number of 6V6s for sale at £4 to £15 each, and 6SK7s for £2 to £12 each.
[3] http://www.sm7ucz.se/Paraset/Paraset_e.htm
[4] Peter R Jensen, *Wireless at War*, pp.175-188, 334-335
[5] Serguief, p.175
[6] n[3] pp. 183-185.
[7] A suggestion by Dan Peterson, W7OIL.

5

21st Century Versions of the Paraset

THOUGH THE PURIST might scoff, there have been attempts by enthusiasts to build Parasets using modern components, such as transistors and ICs. But, all the new versions retain a few key features of the original Paraset, thus maintaining the minimalist design approach to radio building.

First, these modern versions use a regenerative circuit, instead of the heterodyne or direct conversion receiver design found in most modern QRP radios. Of course, everything is processed in analogue circuitry, without digital signal processing. Second, the transmitting frequency is controlled by a crystal oscillator, not by a VFO. Third, there is no built-in speaker, only provision for headphones or earphones. Fourth, most, if not all, employ a straight built-in key on the front panel, just like the original.

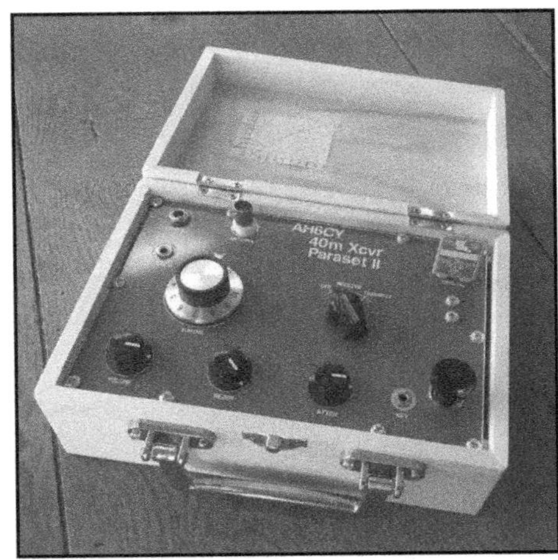

Pic 5.1 Bayou Jumper Paraset by the Four State QRP Group

Self-Build Kits

There are two modern-version Paraset kits on the market at the time of writing:

1. The Bayou Jumper offered by the Four State QRP Group (Pic 5.1)

I built one, and it performs much like my valve version of the Paraset replica, only the Bayou Jumper is more stable. Operating on the 40m band only, there is no CW chirping, and like the original, the key is built into the front panel. The circuit schematics for the receiver and transmitter are shown in Pics 5.2 & 5.3 respectively.

http://4sqrp.com/kits/Bayou%20Jumper/bayoujumper.php

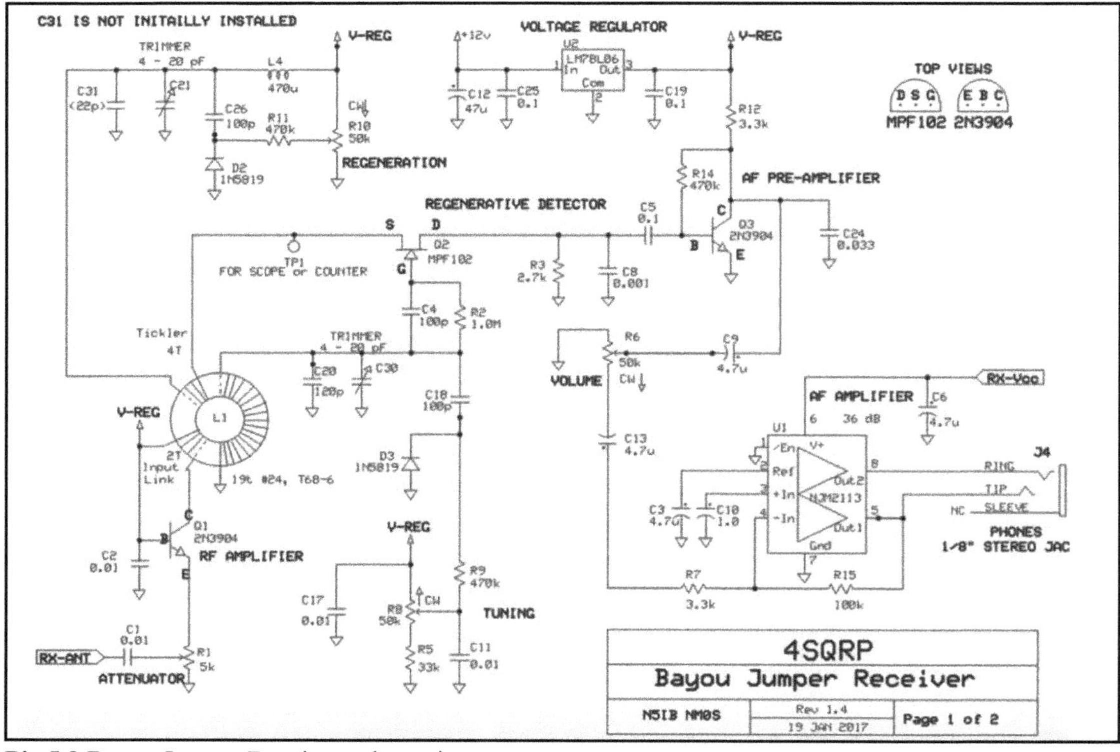

Pic 5.2 Bayou Jumper Receiver schematic

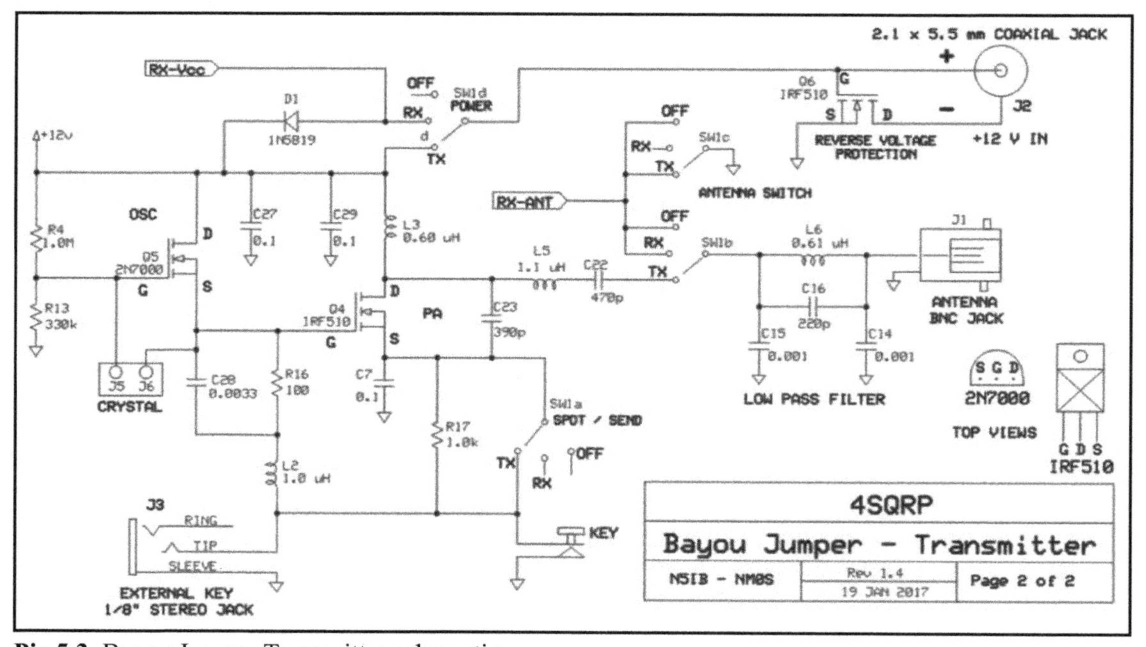

Pic 5.3 Bayou Jumper Transmitter schematic

2. The Para40set (for the 40-meter band) and Para80set (for the 80-meter band). These kits are offered by QRPguys.com of San Jose, California. A photo of the circuit board and its schematics are shown in Pics 5.4 & 5.5 respectively.

I have built and used them on the air. They also are more stable than the original. The key is not built-in with this kit, so you need to supply your own key, which connects externally.

Pic 5.4 The Para80set in enclosure

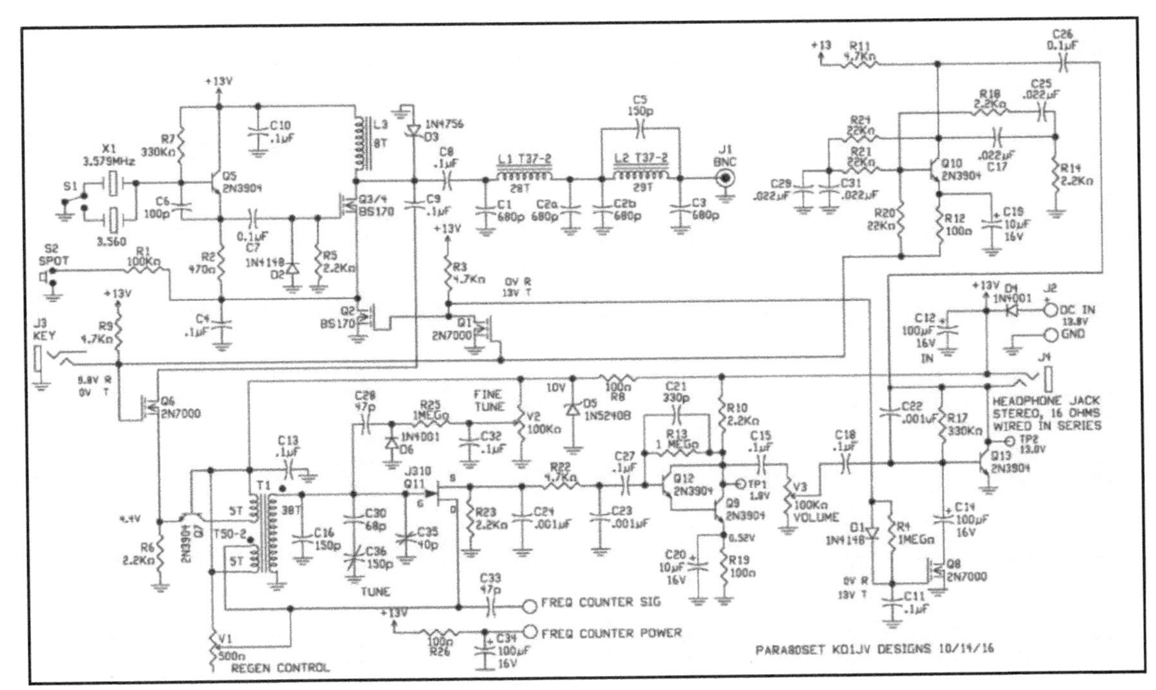

Pic 5.5 The circuit schematic of the Para80set

I have modified the Para80set to receive and transmit on the 40m, as well as the 80m amateur bands, and built it with the same front panel as the original valve Paraset, with unconnected valves as decoration, making the radio look like the original (Pic 5.6). With a 5-pole, 3-position rotary switch, it can be switched between the two sets of L/C tank circuits for both bands, and the required receiving coil/capacitor combinations of the receiver front end

Inside, it looks like this (Pic 5.7):

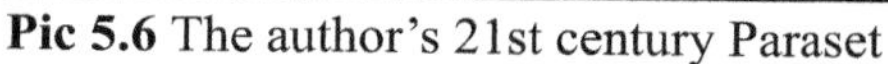

Pic 5.6 The author's 21st century Paraset

Pic 5.7 Inside the Paraset

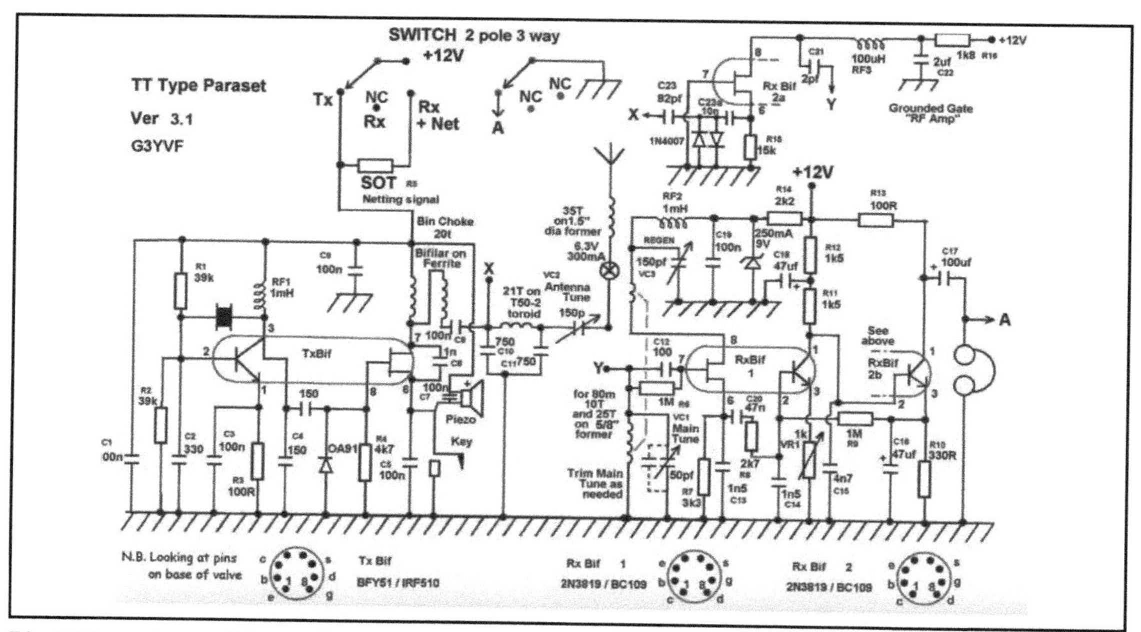

Pic 5.8 Ingenious solid-state Paraset

Further Inspiration

There is one ingenious modern-day Paraset design that retains the look and feel of the original. Built by Geoff, G3YVF, it has hollowed out the metal valves and inserted their transistor equivalents inside. His circuit schematic is shown in Pic 5.8.

Finally, I cannot help but think there must be other 21st Century versions of the Paraset that some creative people have built out there. I would very much like to hear from the builders of such replicas so that I can include them in the next edition of this book.

6

On the air with the Paraset

THE PARASET WORKS as a practical QRP radio on the 40m and 80m Amateur bands. It is not the most stable or easy radio to operate, but many Amateurs who have put their own replicas on the air find the challenge an exhilarating experience. Many mention that they felt as though they were re-living history. That was certainly the case for me. Putting the Paraset on the air is akin to piloting a home-brew vintage bi-plane in today's skies. Some of us, including myself, regularly bring out our Paraset for the ARRL's 'Straight Key Night' [1] on every New Year's Day. The characteristic chirping sound from keying a single-valve oscillator/final valve adds to the romance of operating this spy-set. Note however, that the Paraset was neither designed for a Paraset-to-Paraset QSO nor for co-channel operation that is, both sending and receiving on the same frequency. But with some effort and ingenuity, it is possible to use a Paraset to communicate with other stations on the air just as one would do with a modern transceiver. Many Amateurs have done it, and continue to do so, including making Paraset-to-Paraset QSOs.

Memorable QSOs

One of my fondest memories is the time that I used my Paraset to simulate a 'real world' clandestine operation in November, 2012. I was at the Jalama Beach State Park, 50 miles north of Santa Barbara, California, for a camping and outdoor QRP trip with two of my QRP club buddies. Using a 20-foot-long thin wire flung over a nearby bush for an antenna, I called CQ blindly on 7199.7 kHz. W7UDA in Barstow, California, about 200 miles away, responded and we had a solid two-way QSO. The distance covered would have been more than enough for an SOE agent to send a message from Normandy to the English coast. Of course, thankfully, no Nazis came looking for me! Pic 6.1 was taken by

Pic 6.1 The author operating the Paraset

Pic 6.2 Mike Murphy's Paraset

my good friend Dave Flack, W6DLF (now SK) during that particular QSO.

Two things could be noted about that contact. First, the antenna wire was only 20 feet long for that QSO, far from an ideal length. SOE agents were told to put out a wire antenna as long as 70 feet when possible for better transmission, and to run a random-length counterpoise wire attached to the "E" (ground) connector. The Paraset tuning circuit design allows the operator to tune practically any length of wire. Second, I was operating very close to the seawater's edge and in an area where there was no stray RF noise. (There was not even a cell-phone nor WiFi signal in the park then). No one had a TV operating, either. In other words, I was at an ideal radio location. As many Amateurs know, when you operate your radio near a large body of salt water, you have a "sea water amp" effect. Your 5 watts signal can radiate almost as though it were 500 watts. Such conditions probably did not exist for most clandestine users in World War II.

Mike Murphy, WU2D, whose excellent Paraset reproduction is shown here (Pic 6.2), had a most extraordinary QSO using his set (Pic 6.3). He related the following story to me:

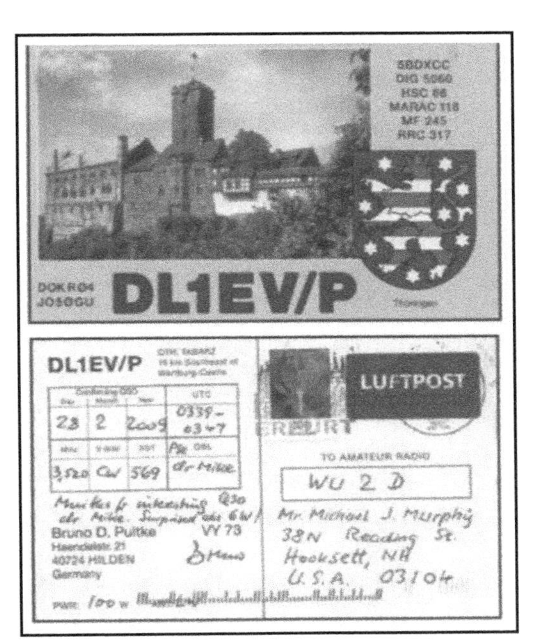

Pic 6.3 DL1EV's QSL card

Pic 6.4 Steve McDonald's Paraset

"…[O]n Feb. 27, 2009…I boldly called a CQ on 3520 kHz at 10:30 p.m. and DL1EV came back from Germany… We QSO'd for about 20 minutes. It turned out he was a radio operator in the Wehrmacht in WWII and actually hunted Parasets with his DF (direction-finding) van."

DX Contacts

Steve McDonald, VE7SL, holds what may be the distance record for a Paraset QSO. His contact spanned the distance between British Columbia (Canada), and Cape Town (South Africa), which are over 10,000 miles apart. Steve's Paraset reproduction is shown in Pic 6.4, and the QSL card attesting to this remarkable achievement in Pic 6.5

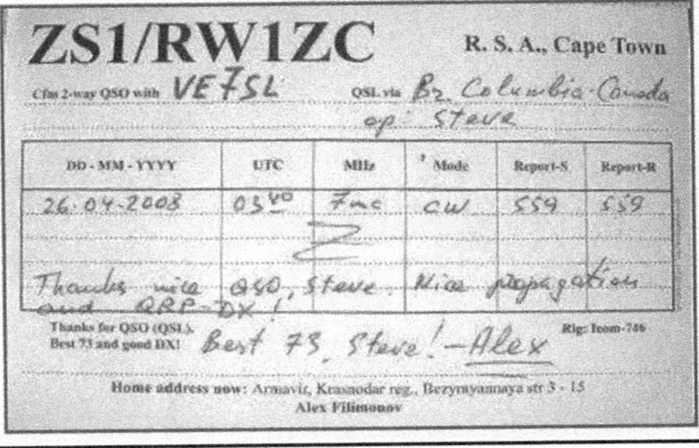

Pic 6.5 QSL card received by Steve VE7SL from Alex ZS1/RW1ZC

On Steve's website (https://qsl.net/ve7sl/paraset.html), he shows his collection of QSL cards for contacts made with his Paraset.

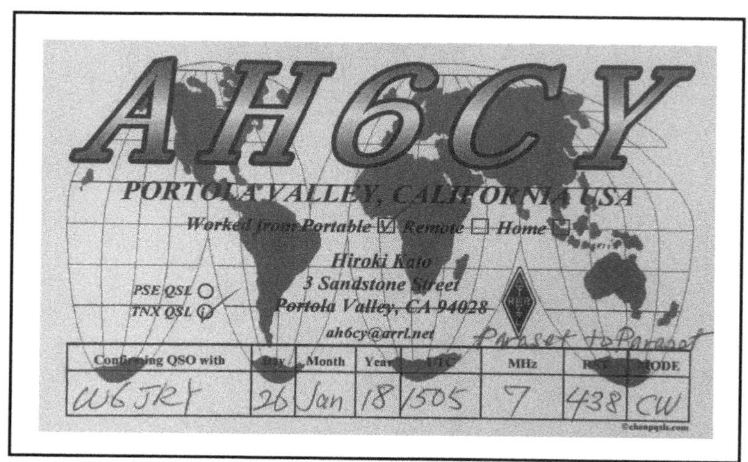

Pic 6.6 The author's QSL card

Skeds

Normally, a Paraset-to-Paraset contact takes place through a 'sked' (or schedule) arrangement as one doesn't often encounter Paraset operators given the small number of Paraset radios out there, including originals and reproductions,. I had a Paraset-to-Paraset QSO with Jerry Fuller, W6JRY, on the 40m band via a sked (Pic 6.6). We live about 200 miles apart, Jerry in Forest Ranch, California, and me in Portola Valley, California.

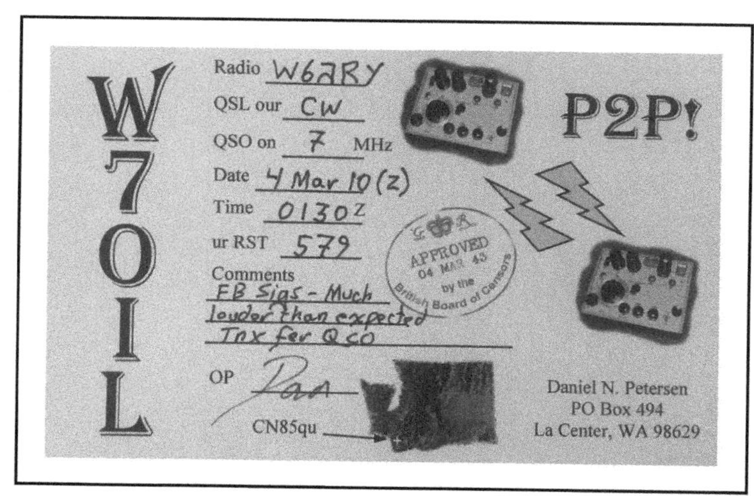

Pic 6.7 P2P QSL card

QSL Cards

Dan Petersen, W7OIL, issues a special P2P (Paraset-to-Paraset) contact QSL card. Dan's QTH is in Washington State. He has had P2P contacts with Jerry Fuller, W6JRY, in California; Steve MacDonald, VE7SL, in British Columbia (Canada); and Paul Signorelli, W0RW, in Colorado. His farthest P2 non-P QSO was with a station in Michigan.

The Future

Though the Paraset-to-Paraset QSO is a rarity at the moment, I am not without hope that some day soon we will see more Amateurs on the air with their Paraset reproductions as it becomes better known and better appreciated. There may soon be a Paraset net operation day on a regular basis. I am saying this because I see a few encouraging signs towards that goal as I write: there are an increasing number of Amateurs trying out all manner of QRP rigs, commercially built, home-brewed, and some built from kits. As we saw in the last chapter, there are at present two purveyors of the 21st Century versions of the Paraset kits (4SQRP.com and QRPguys.com). For the purist, these may not be as appealing as a valve set, but those who get their appetite whetted by building and using these sets on the air would likely be interested in going further 'back to the future', especially if someone offers a complete kit with valves and reproduction 'vintage' parts. There may be a burgeoning market. If you look at the modern audio market, you will notice many inexpensive valve amps, both plug-and-play and in kit form, are increasingly coming from China and other countries bringing back the fun of valve audio. Given that many QRP transceiver kits are also being exported from China and Russia, it wouldn't be surprising if we were to begin to see the original-design valve Paraset in kit form coming from some of the same sources.

Pic 6.8 Jerry Fuller's spotting device

As the Paraset was intended and designed for split-frequency operation (receiving and transmitting on different frequencies) and for cross-mode use (listening to BBC AM broadcast and transmitting in the CW mode), it did not have any 'spotting' capability, that is, the receiver cannot readily be tuned to the transmitting frequency precisely. Some Amateurs, like, Jerry W6JRY, have built a separate stand-alone crystal oscillator and tune the Paraset's receiver to that frequency (Pic 6.8). I use another crystal-controlled QRP transmitter connected to a dummy load and generate a very small signal to tune the Paraset's receiver. It remains for someone to incorporate a simple spotting circuit into a modern Paraset replica, enhanced by that newer feature.

As we close this book, it is my hope that the readers will forward stories of their own experiences with the Paraset and its modern iterations together with comments and additional information to improve this book for future editions. It is also my hope that many more people will be building their own Parasets, whether with valves or transistors, to keep the legacy of this unique radio alive.

See you on the air with your Paraset!

Reference

[1] Straight Key Night is a 24-hour event every New Year's Day, during which Amateurs in North America get on the air using a straight (manual telegraph) key to celebrate the legacy of Morse code and old radios. Many Amateurs dust off old valve radios in order to participate in the event.

Appendix 1: The Coding Scheme used by SOE

Reproduced from M.R.D. Foot's *SOE: The Special Operations Executive 1940-46* pp.122-124. Explanatory comments added below (in brackets);

(A coding key is produced from any memorable line of verse)

Let us take an easy example: 'Who killed Cock Robin?' It is written out in lines of five letters, omitting any letters used already; the rest of the alphabet then fills the 5 x 5 letter square. I and J count as one.

W H O K IJ
L E D C R
B N A F G
M P Q S T
U V X Y Z

The message to be sent - let us suppose another easy one, 'Robert taken' - is divided into bigrams, groups of two letters; any dud being used to fill up a blank space; RO BE RT TA KE NY. Each of these bigrams is encoded by taking the two opposite corners of the rectangle it forms in the word square; thus RO becomes DI or DJ. If both letters of the bigram are in the same line, the next letter to the right of each is used; if both are in the same column, the next letter below. RT thus become GZ. The simple message, simply encoded, becomes DI NL GZ QG HC FV. The wireless operator can confuse things slightly more, by putting the bigrams in groups of five letters: DINLG ZQGHC FV, and filling up the gap at the end with three more null letters. This is by no means an impenetrable code.....though it takes a clever intelligence officer with a particular cast of mind to unravel a Playfair quickly. It might provide worse degree of tactical security for a circuit.

(SOE banned the use of Playfair in 1942 and in its place introduced a system called 'double transposition')

For this the agent had to remember two random numbers, each some six or seven figures long;

(Agents were not supposed to write them down - but those who did not trust their memory did so to avoid forgetting them)

The plain text was written out under the first random number; then transcribed, reading each column of letters vertically downwards, in the numerical order off each column; then, so enciphered, written down under the second random number, and again read off by numerical order off columns (3 before 4 and so on). Again, an illustration may help. Let us take 487245 and 3258497 as the random numbers, and 'Robert taken Friday' as the message. First, the message is written out under the first number (below left). The columns are then read off in numerical order: EEA RTR TF BKD OAI RNY. This text is then written out under the second number (below right):

4	8	7	2	9	5
R	O	B	E	R	T
T	A	K	E	N	F
R	I	D	A	Y	

3	2	5	8	4	9	7
E	E	A	R	T	R	T
F	B	K	D	O	A	I
R	N	Y				

Again the columns are read off in numerical order: EBN EFR TO AKY TI RD RA, and the result if put in groups of five letters: EBNEF RTOAK YTIRD RA. This text is ready for transmission. At the receiving end, the text is written out in vertical columns, by numerical order of column again, under the secondly, read off horizontally, and put under the first key in vertical columns; the message then ought to be readable horizontally at a glance.

(There were further revisions and improvements made until the SOE settled on the one-time pad)

Appendix 2: Surviving Original Parasets

Global list of surviving original Parasets, as of Feb 9, 2019, compiled by AH6CY and F6EJU.

Parasets in Museums:

Musée de la Résistance Bretonne, France (x3)

Musée de la Résistance de Vassieux en Vercors, France

Musée Mémorial des Combats de la Poche de Colmar,

Turckheim, Alsace, France

Musée 'MMPark' de la Wantzenau, Strasbourg, France (SN#7694)

Bletchley Park, Milton Keynes UK (x2) (owned by David White, no longer on display)

Imperial War Museum, London, UK

Military Wireless Museum, Kidderminster, UK (SN# 2571)

National Military Museum, Soest, The Netherlands

War Museum Overloon, Overloon, The Netherlands (SN# 10529)

Norges Hjemmefronmuseum (Norwegian Resistance Museum), Oslo, Norway

Parasets in Private Collections:

Private collection owned by Mark Meltzer, AF6IM (SN#7428)

Private collection owned by Patrick, F4SMX (SN#7629)

Private collection owned by Pascal Drouvin, F8JZR (SN#7649)

Private collection owned by Bernard Van Haecke, KI6TSF (radio located in Belgium)

Private collection owned by Geet Willendrup, Denmark (SN#2357)

Private collection owned by David White, G3ZPA

Private collection owned by Alan Oately, M0AVN (SN#10448)

Private collection owned by Helge Fykse Helge, LA6NCA (SN# 7606)

Private collection owned by PA0SE

Key: SN# = Paraset serial number

B1

Bibliography

Atkin M, *Fighting Nazi Occupation: British Resistance 1939-1945* (Pen & Sword, Barnsley 2015)

Basu S, *Spy Princess: The Life of Noor Inayat Khan*, (Sheridan Books, Michigan 2007)

Cookridge EH, *Inside SOE* (Heinemann, London 1966)

Cornioley PW, *Code Name Pauline: Memoirs of a World War II Special Agent*, (Chicago Review Press 2013)

Cruickshank C, *SOE in Scandinavia*, (OUP 1986)

Escott BE, *The Heroines of SOE: Britain's Secret Women in France F Section*, (The History Press, Stroud 2012)

Follett K *Jackdaws*, (Signet, New York 2002)

Foot MRD, *SOE: The Special Operations Executive 1940-46* (BBC, London, 1984)

_________, *SOE in France* (Whitehall History Publishing, London, 2004)

Giskes HJ, *London Calling North Pole: The True Revelations of a German Spy* (Echo Point Books & Media, Brattleboro 1953). Translated from original German

Gordon KG, 'The "Paraset" Suitcase Spy Transceiver of WWII' (*Glowbugs.* W7EKB;

Archived from the original on 2006. Retrieved 5 July 2016)

Grahan J, *Unearthing Churchill's Secret Army: The Official List of SOE Casualties and Their Stories* (Pen & Sword Books, Barnsley 2012)

Guiet, C, Smith TK, *Scholars Of Mayhem: My Father's Secret War in Nazi-Occupied France* (Penguin Press, New York 2019)

Hudson S, *Undercover Operator: An SOE Agent's Experiences in France & the Far East* (Leo Cooper, Barnsley 2003)

Jensen PR, *Wireless at War, Developments in Military and Clandestine Radio 1895-2012* (Australia, 2013)

Krame R, *Flames in the Field: The Story of Four SOE Agents in Occupied France,* (Penguin Books, London 1995)

Kato H. "The Paraset: A WWII Spy Radio You Can Build," *CQ* Feb. 2016, pp 19-25

__________. 'Two Clandestine Radio of WWII: Replicating the Prison Camp Radio and the Paraset Spy Transceiver', *(Electric Radio* Nov. 2012, pp 26-39)

__________. 'Two Clandestine Radios of WWII- Update', (*Electric Radio* March 13, pp 18-26)

Korbonski S, *Fighting Warsaw: The Story of the Polish Underground State 1939-45* (George Allen & Unwin, London 1956)

Ladd J, Melton K, *Clandestine Warfare: Weapons and Equipment of the SOE and OSS* (Blandford Press, London 1988)

Loftis L, *Code Name: Lise: The True Story of the Woman Who Became WWII's Most Highly Decorated Spy* (Gallery Books, New York 2019)

Mackenzie W, *The Secret History of Special Operations Executive 1940-1945* (2nd edn, St Ermin's Press, London 2002)

Mace M, Grehan J, *Unearthing Churchill's Secret Army: The Official List of SOE Casualties and Their Stories* (Pen & Sword, Barnsley 2012)

McClain S, *Navajo Weapon: The Navajo Code Talkers*, (Rio Nuevo Publishers, Tucson 2001)

Meulstee L, Staritz RF, *Wireless For The Warrior, Vol. 4 Clandestine Radio* (Wimborne Publishing Limited, Ferndown 2004)

Ottaway S, *Violetter Szabo: the Life that I have* (Thistle Publishing, London 2002)

Pearson, JL, *The Wolves at the Door: The True Story of America's Greatest Female Spy* (The Lyons Press, Guildford, 2005)

Purnell S, *A Woman of No Importance: The Untold Story of the American Spy Who Helped Win World War II* (Viking, 2019)

Perqiun, J-L, *The Clandestine Radio Operators*, (Historie & Collections, Paris, 2011)

__________, *Clandestine Parachute and Pick-Up Operations, Vol. 1*, (Historie & Collections, Paris, 2012)

Pidgeon G, *The Secret Wireless War: The Story of MI6 Communications 1939-1945* (2nd edn, St Leonards-on-Sea, UPSO 2007).

Reed O, Oluf, *Two Eggs On My Plate* (George Allen & Unwin, London 1952). Translated from the Norwegian by F. H. Lyon.

Riols N, *The Secret Ministry Of Ag. & Fish* (Pan Macmillan, London 2013).

Rose S, *D-Day Girls: The Spies Who Armed the Resistance, Sabotaged the Nazis, and Helped Win World War II* (Crown, 2019)

Sergueiev L, *Secret Service Rendered* (William Kimber & Co, London 1966)

Stevenson, W, *Spymistress* (Arcade Publishing, New York 2011)

The National Archives, *SOE Manual: How to be an Agent in Occupied Europe* (William Collins, London 2014)

Verity H *We Landed By Moonlight: Secret RAF Landings in France, 1940-1944* (Crecy Publishing, Manchester1978)

References

The author wishes to extend his warmest appreciation to the following organizations and individuals for permission to use their materials for this book.

The Imperial War Museum, Pics 2.1, 2.2, 2.6, 4.14

The Bomber Command Museum of Canada, Pic 2.3

The Crypto Museum, Pic 2.5

Jeffrey Bass, Pic 2.8

Louis Meulstee, Pics 3.1-3.4, 3.6-3.9 Figs 3.1-3.5

Peter Jensen, VK7AKJ, Pics 4.17, 4.62

Jean-Claude Moffet, F6EJU, Pics 4.16, 4.25-4.27

Henk van Zwam, Pics 4.22, 4.23, 4.35

Johnny Appel, SM7UCZ, Pics 4.6-4.10, 4.32

Daniel Peterson, W7OIL, Pics 4.28, 2.63, 6.7

Steve McDonald, VE7SL, Pics 4.46, 6.5

Richard Bonkowski, W3HWJ, Pics 4.6, 4.58

Four State QRP Group, Pic 5.2, 5.3

QRPGuys, Pic 5.5

Geoffrey Wooster, G3YVF, Pic 5.8

Mike Murphy, WU2D, Pics 6.2-6.4